An Introduction to Mycosporine-Like Amino Acids

Authored by

Hakuto Kageyama

Graduate School of Environmental and Human Sciences/
Faculty of Science and Technology
Meijo University
Japan

An Introduction to Mycosporine-Like Amino Acids

Author: Hakuto Kageyama

ISBN (Online): 978-981-5136-08-1

ISBN (Print): 978-981-5136-09-8

ISBN (Paperback): 978-981-5136-10-4

Published by Bentham Science Publishers Pte. Ltd. Singapore.

First published in 2023.

BENTHAM SCIENCE PUBLISHERS LTD.

End User License Agreement (for non-institutional, personal use)

This is an agreement between you and Bentham Science Publishers Ltd. Please read this License Agreement carefully before using the ebook/echapter/ejournal (**"Work"**). Your use of the Work constitutes your agreement to the terms and conditions set forth in this License Agreement. If you do not agree to these terms and conditions then you should not use the Work.

Bentham Science Publishers agrees to grant you a non-exclusive, non-transferable limited license to use the Work subject to and in accordance with the following terms and conditions. This License Agreement is for non-library, personal use only. For a library / institutional / multi user license in respect of the Work, please contact: permission@benthamscience.net.

Usage Rules:

1. All rights reserved: The Work is the subject of copyright and Bentham Science Publishers either owns the Work (and the copyright in it) or is licensed to distribute the Work. You shall not copy, reproduce, modify, remove, delete, augment, add to, publish, transmit, sell, resell, create derivative works from, or in any way exploit the Work or make the Work available for others to do any of the same, in any form or by any means, in whole or in part, in each case without the prior written permission of Bentham Science Publishers, unless stated otherwise in this License Agreement.
2. You may download a copy of the Work on one occasion to one personal computer (including tablet, laptop, desktop, or other such devices). You may make one back-up copy of the Work to avoid losing it.
3. The unauthorised use or distribution of copyrighted or other proprietary content is illegal and could subject you to liability for substantial money damages. You will be liable for any damage resulting from your misuse of the Work or any violation of this License Agreement, including any infringement by you of copyrights or proprietary rights.

Disclaimer:

Bentham Science Publishers does not guarantee that the information in the Work is error-free, or warrant that it will meet your requirements or that access to the Work will be uninterrupted or error-free. The Work is provided "as is" without warranty of any kind, either express or implied or statutory, including, without limitation, implied warranties of merchantability and fitness for a particular purpose. The entire risk as to the results and performance of the Work is assumed by you. No responsibility is assumed by Bentham Science Publishers, its staff, editors and/or authors for any injury and/or damage to persons or property as a matter of products liability, negligence or otherwise, or from any use or operation of any methods, products instruction, advertisements or ideas contained in the Work.

Limitation of Liability:

In no event will Bentham Science Publishers, its staff, editors and/or authors, be liable for any damages, including, without limitation, special, incidental and/or consequential damages and/or damages for lost data and/or profits arising out of (whether directly or indirectly) the use or inability to use the Work. The entire liability of Bentham Science Publishers shall be limited to the amount actually paid by you for the Work.

General:

1. Any dispute or claim arising out of or in connection with this License Agreement or the Work (including non-contractual disputes or claims) will be governed by and construed in accordance with the laws of Singapore. Each party agrees that the courts of the state of Singapore shall have exclusive jurisdiction to settle any dispute or claim arising out of or in connection with this License Agreement or the Work (including non-contractual disputes or claims).
2. Your rights under this License Agreement will automatically terminate without notice and without the

Bentham Science Publishers Pte. Ltd.
80 Robinson Road #02-00
Singapore 068898
Singapore
Email: subscriptions@benthamscience.net

CONTENTS

FOREWORD

This book is a good introduction to those interested in Mycosporine-like amino acids (MAAs), and will be an asset for students, university researchers, and corporate researchers.

MAAs are natural compounds synthesized by microorganisms to protect themselves from UV irradiation. Their usefulness is expected in various biological and industrial fields. This book describes MAAs in detail, from the basics to the applied perspectives. This book is composed of 11 sections. First, starting with the molecular structure of MAAs (Chapter 1), their distributions are outlined, focusing on cyanobacteria (Chapter 2). Chapter 3 has a detailed description of the MAAs biosynthetic pathway. These chapters will help readers gain a basic understanding of MAAs. Chapter 4 introduces the knowledge accumulated so far regarding the method of analysis and preparation of MAAs. The characteristics of MAAs are described in chapter 5 to 11. Each chapter is compactly organized from the point of view application so that readers will find the usefulness of MAAs. It should be noted that this book has abundant appendices. Information on more than 60 types of MAAs reported so far is summarized. Correlation diagrams of the molecular structures of these many MAAs are also available. This book thus can serve as a key reference work to all those working on MAAs.

Rungaroon Waditee-Sirisattha, Ph.D.
Department of Microbiology, Faculty of Science
Chulalongkorn University
Thailand

PREFACE

Mycosporine-like amino acids (MAAs), which have the property of absorbing ultraviolet rays, are natural compounds that are applied in the cosmetics field as the active ingredient in sunscreens. In recent years, it has become clear that MAAs have various useful functions such as antioxidant activity and anti-inflammatory action as well as ultraviolet absorption ability. Therefore, applications may be considered and developed not only in the cosmetics field but also in various fields such as pharmaceuticals and foods. Patents have already been filed by companies around the world, and cosmetic ingredients and skin care products containing MAAs have been launched on the market.

This book details the molecular structures, activities, and application examples of MAAs and is intended to be used as an introductory book for undergraduate and graduate students in science or as a handbook for researchers. I aim to make the descriptions as clear as possible. In the text, references to academic research are cited as appropriate. I hope that this book will help to deepen the knowledge of MAAs.

This book is based on an English translation of my book Mycosporine-like Amino Acids Nyumon, published by Sankeisha in Japan in 2021. I was able to take the opportunity to review and remake the full text of the book and to incorporate the latest information. I would like to thank all the people involved in the production of this book. I would also like to take this opportunity to thank Dr. Rungaroon Waditee-Sirisattha of Chulalongkorn University, a collaborator who has been conducting research on cyanobacteria and MAAs for many years.

Hakuto Kageyama
Graduate School of Environmental and Human Sciences/
Faculty of Science and Technology
Meijo University
Japan

CHAPTER 1

Mycosporine-like Amino Acids and their Biomolecular Properties

Abstract: Mycosporine-like amino acids (MAAs) are natural ultraviolet (UV)-absorbing compounds that are attracting attention in the industrial field including cosmetics and pharmaceuticals. This book provides a wide range of descriptions of MAAs, from fundamentals to applications. In order to discuss the properties of MAAs, an understanding of their chemical structures would be required. The purpose of this chapter is to understand the basic molecular structure of MAAs. In general, MAAs have structures in which amino acids are bound to the core structures of cyclohexenone or cyclohexenimine. In addition to the basic structure, the resonance hybrid structures of MAAs are also described here. Delocalization of electrons is considered to affect the stability and the absorption maximum wavelength of MAA molecules. We will also discuss the environmental factors that can affect the structure of MAAs. Finally, databases of molecular structure information of MAAs will be described.

Keywords: Cyclohexenimine, Cyclohexenone, Environmental factors, Maximum absorption wavelength, Molar absorption coefficient, Molecular structure, Mycosporine-like amino acid, Resonance hybrid structure, Ultraviolet.

INTRODUCTION

Mycosporine-like amino acids (MAAs) are water-soluble small organic compounds containing nitrogen in their molecular structure and are known as natural sunscreens. "Mycosporine" is a secondary metabolite that originally existed in fungi. It has a specific molecular structure in which amino acids bound to its basic structure are called MAAs [1]. To explain in a little more detail, it is as follows. It has long been known that fungi have substances with maximum absorption in the UV region (310 nm). Since this substance was thought to be involved in sporulation, it was called mycosporine (myco- + spore + -ine) together with the prefix myco-, which means fungi. In 1976, one of the chemical structures of mycosporine was reported (currently this substance is called mycosporine-serinol) [2]. After that, it became clear that compounds with similar structures also exist in various organisms other than fungi. Since amino acids were contained in the molecular structure, these compounds came to be called mycosp-

Hakuto Kageyama

orine-like amino acids (hereinafter referred to as MAA). To date, more than 60 kinds of MAA compounds have been reported.

MAAs are well-known UV-absorbing compounds. The maximum absorption wavelengths of MAAs are in the range of 310 to 362 nm. In addition, the value of the molar absorption coefficient is as large as $\varepsilon = 20{,}000$ to $50{,}000\ M^{-1}\ cm^{-1}$. As an example, the absorption spectrum of mycosporine-2-glycine, which is a type of MAA purified from salt-tolerant cyanobacteria, is shown in (Fig. **1**). Given that UV rays are classified into UV-A (315–400 nm), UV-B (280–315 nm), and UV-C (100–280 nm) according to wavelength, MAAs are substances that efficiently absorb UV-A and UV-B. They are also considered to be compounds with the strongest UV-A absorption capacity in nature [3]. MAAs can release the absorbed UV energy to the surroundings as heat without producing harmful substances such as reactive oxygen species (ROS) [4].

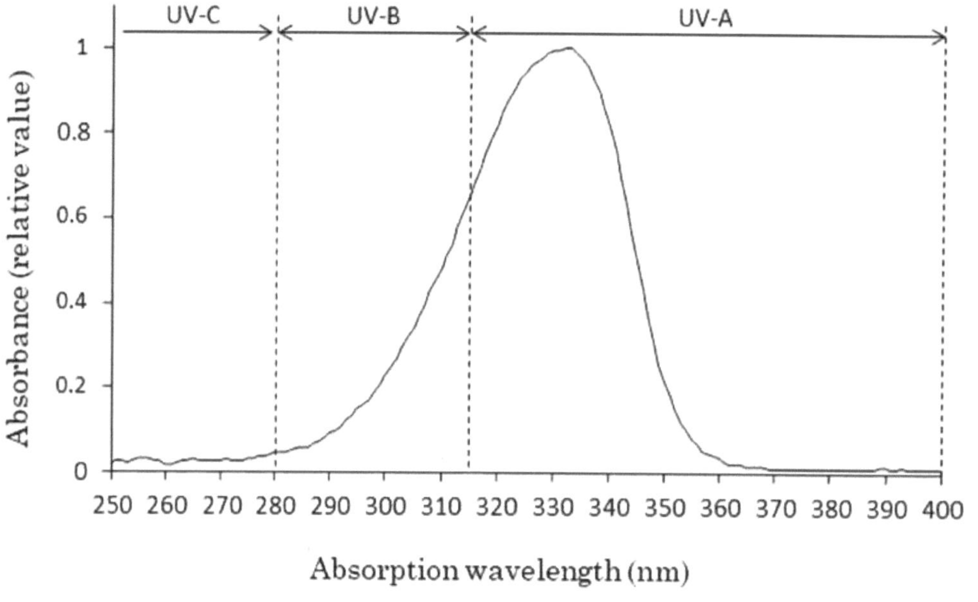

Fig. (1). Absorption spectrum of an MAA (mycosporine-2-glycine).

Many species that biosynthesize MAAs have been reported. So far, MAAs are found in various marine, freshwater, and terrestrial species, including micro and macroalgae, cyanobacteria, and animals [5 - 9]. It is thought that MAAs accumulated in the organism contribute to the reduction of damage to nucleic acids and proteins caused by UV rays. In addition, as will be described later, various physiological activities other than UV absorption have been reported.

MOLECULAR STRUCTURES OF MAAS

Basic Chemical Structure

The core part of the molecular structure of MAAs is a cyclohexenone structure or cyclohexenimine structure (Fig. **2**) [10, 11]. Basically, the substituted amino acids are bound as moieties R_1 and R_2. MAAs with a cyclohexenone structure contain one amino acid, and when R_1 is replaced with glycine, it produces mycosporine-glycine (Fig. **3**). However, in the cyclohexenimine structure, two substituents are substituted. For example, when R_1 and R_2 are substituted with glycine and serine, respectively, they produce shinorine (Fig. **3**). In disubstituted MAAs, the amino acid corresponding to the R_1 moiety is often glycine. This is because glycine first binds to the basic structure to produce mycosporine-glycine, and then the second amino acid binds to mycosporine-glycine to form disubstituted MAAs in the MAA biosynthetic pathway. (Details of MAA biosynthetic pathways will be described in Chapter 3.) Table **1** shows the molecular structure, substituents, and absorption maxima of representative MAAs.

Cyclohexenone Cyclohexenimin

Fig. (2). Core structures of MAAs.

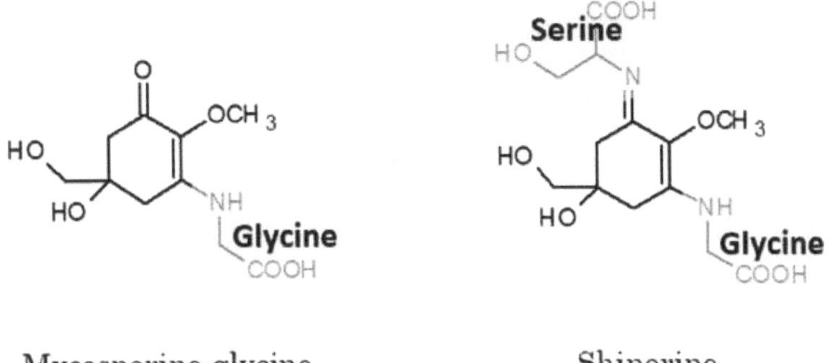

Fig. (3). Amino acid substitutions of core structures.

Table 1. Substituents for representitive MAAs.

MAA	Substituent R_1	Substituent R_2	Absorption Maximum (nm)
Monosubstituted MAA	-	-	-
Mycosporine-glycine	Glycine	-	310
Mycosporine-GABA	GABA	-	310
Disubstituted MAA	-	-	-
Shinorine	Glycine	Serine	334
Porphyra-334	Glycine	Threonine	334
Mycosporine-2-glycine	Glycine	Glycine	332
Palythine	Glycine	Amino group ($-NH_2$)	320

(Table 1) cont.....

MAA	Substituent R_1	Substituent R_2	Absorption Maximum (nm)
Palythine-serine 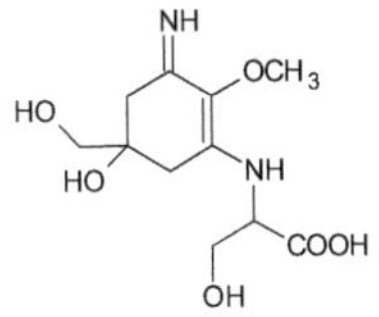	Serine	Amino group ($-NH_2$)	320
Euhalothece-362 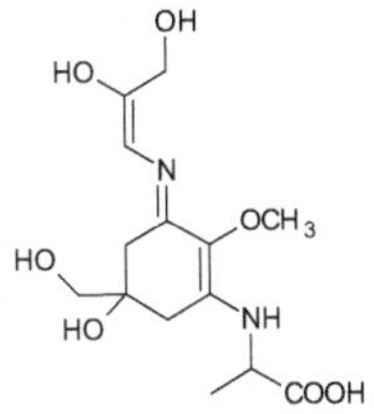	Alanine	2,3-Dihydroxypropenyl amine	362
Asterina-330 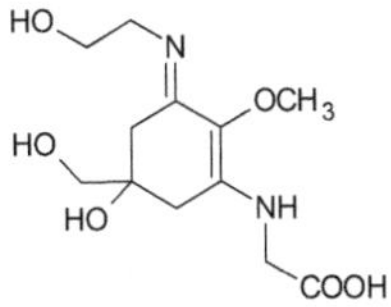	Glycine	2-Aminoethanol	330
Palythene 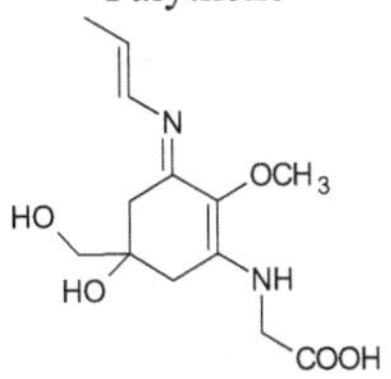	Glycine	1-Aminopropene	360

Resonance Hybrid Structures

MAAs exist as zwitterions with both positive and negative charges in one molecule. As shown in (Fig. **4**), a resonance hybrid structure is formed by the superposition of two types of canonical structures. Delocalization of a positive charge occurs between nitrogen atoms and remains sandwiched between the C1 to C3 positions of the ring. Such a structure is called a conjugated structure. It is considered that MAA molecules are stabilized by lowering the energy level due to the delocalization of electrons. In addition, the conjugated structure is known to affect the wavelength of light absorbed by the molecule. Therefore, it is considered that the degree of delocalization contributes to the absorption maximum wavelength and the value of the molar absorption coefficient of each MAA. Generally, the longer the conjugated structure, the longer the wavelength of the light absorbed.

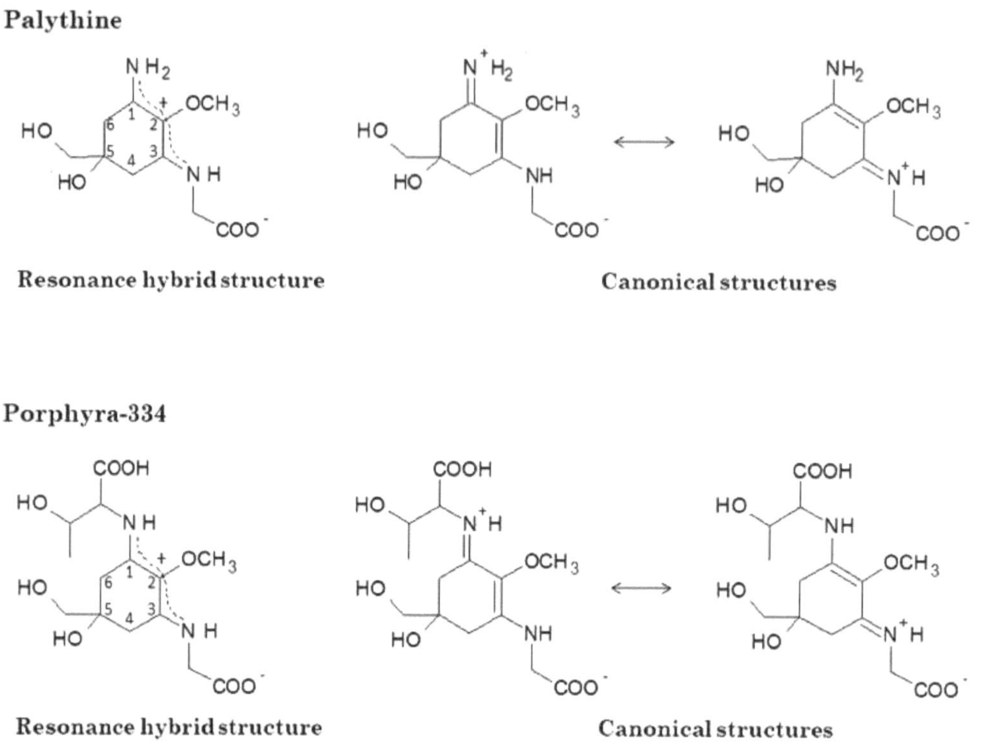

Palythine

Resonance hybrid structure

Canonical structures

Porphyra-334

Resonance hybrid structure

Canonical structures

Fig. (4). Resonance hybrid structures of palythine and porphyra-334.

Factors that Affect the Molecular Structures of MAAs

It has been reported that the pH and temperature of the solvent affect the structures of MAAs. The maximum absorption wavelength of porphyra-334 dissolved in aqueous solution with a pH near neutral was 334 nm, but it changed to 332 nm at pH = 3 and 330 nm at pH = 1–2 [12]. It is believed that excess protons in a highly acidic aqueous solution bind to lone pairs present in the nitrogen atom in the porphyra-334 molecule, causing protonation. Thus, the delocalization of positive charges in the conjugated structure might be inhibited by the protonation, resulting in a smaller absorption maximum wavelength. It has also been reported that protonation reduced the absorption maximum wavelength in mycosporine-glycine and shinorine [13]. This report showed that the protonation of the carboxylate anion ($R\text{-}COO^-$) in the amino acid residues contained in MAAs was involved in the change in the absorption maxima. However, it has been shown that under high alkaline conditions (above pH 12), there was no change in the absorption maximum, but the absorbance of porphyra-334 decreased, and another compound with an absorption maximum at 225 nm was produced. These results suggested that porphyra-334 became unstable under strongly alkaline conditions, and its degradation products were formed. In addition, high temperature contributed to the stability of porphyra-334. At above 60°C, the degradation of porphyra-334 was promoted not only in alkaline solutions but also in acidic solutions [12].

DATABASE OF MAAS

A database of MAAs named Mycas has recently been published by Geraldes *et al.* (http://www.cena.usp.br/ernani-pinto-mycas) [14]. This database covers more than 70 MAAs as of February 2022 and contains basic information such as molecular formula, exact mass, and MS fragments of each MAA. This book includes an appendix that lists information on each MAA, including its chemical structure.

CONCLUDING REMARKS

MAAs are promising natural organic compounds that can be applied to the cosmetics and pharmaceutical fields. An understanding of the chemical structure of MAAs is essential for investigating and assessing the usefulness of MAAs. This chapter described the basic molecular structures of MAAs. This short chapter would be important because it underlies the wide range of topics described in this book, such as MAA biosynthetic pathways and their regulatory mechanisms, analytical methods of MAAs, and useful activities of MAAs.

LIST OF ABBREVIATIONS

GABA gamma-aminobutyric acid

MAA mycosporine-like amino acid

ROS reactive oxygen species

UV ultraviolet

CONSENT FOR PUBLICATION

Not applicable.

CONFLICT OF INTEREST

The author declares no conflict of interest, financial or otherwise.

ACKNOWLEDGEMENTS

Declared none.

REFERENCES

[1] Oren, A.; Gunde-Cimerman, N. Mycosporines and mycosporine-like amino acids: UV protectants or multipurpose secondary metabolites? *FEMS Microbiol. Lett.,* **2007**, *269*(1), 1-10. [http://dx.doi.org/10.1111/j.1574-6968.2007.00650.x] [PMID: 17286572]

[2] Favre-Bonvin, J.; Arpin, N.; Brevard, C. Structure de la mycosporine (P 310). *Can. J. Chem.,* **1976**, *54*(7), 1105-1113. [http://dx.doi.org/10.1139/v76-158]

[3] Schmid, D.; Schürch, C.; Zülli, F. *UVA-screening compounds from red algae protect against photoageing*; Personal Care, **2004**, pp. 29-31.

[4] Conde, F.R.; Churio, M.S.; Previtali, C.M. The deactivation pathways of the excited-states of the mycosporine-like amino acids shinorine and porphyra-334 in aqueous solution. *Photochem. Photobiol. Sci.,* **2004**, *3*(10), 960-967. [http://dx.doi.org/10.1039/b405782a] [PMID: 15480487]

[5] Rosic, N. Genome mining as an alternative way for screening the marine organisms for their potential to produce UV-absorbing mycosporine-like amino acid. *Mar. Drugs,* **2022**, *20*(8), 478. [http://dx.doi.org/10.3390/md20080478] [PMID: 35892946]

[6] Wei, Y.; Nishiuchi, T.; Sakamoto, T. Characterization of mycosporine-like amino acids in the edible cyanobacterium Nostoc commune (Di Pi Cai) from China. *J. Gen. Appl. Microbiol.,* **2021**, *67*(6), 260-264. [http://dx.doi.org/10.2323/jgam.2021.03.003] [PMID: 34349076]

[7] Sinha, R.P.; Singh, S.P.; Häder, D.P. Database on mycosporines and mycosporine-like amino acids (MAAs) in fungi, cyanobacteria, macroalgae, phytoplankton and animals. *J. Photochem. Photobiol. B,* **2007**, *89*(1), 29-35. [http://dx.doi.org/10.1016/j.jphotobiol.2007.07.006] [PMID: 17826148]

[8] Figueroa, F.L. Mycosporine-like amino acids from marine resource. *Mar. Drugs,* **2021**, *19*(1), 18. [http://dx.doi.org/10.3390/md19010018] [PMID: 33406728]

[9] Carreto, J.I.; Carignan, M.O. Mycosporine-like amino acids: relevant secondary metabolites. Chemical and ecological aspects. *Mar. Drugs,* **2011**, *9*(3), 387-446. [http://dx.doi.org/10.3390/md9030387] [PMID: 21556168]

[10] Shick, J.M.; Dunlap, W.C. Mycosporine-like amino acids and related Gadusols: Biosynthesis, acumulation, and UV-protective functions in aquatic organisms. *Annu. Rev. Physiol.,* **2002**, *64*(1), 223-262. [http://dx.doi.org/10.1146/annurev.physiol.64.081501.155802] [PMID: 11826269]

[11] Rosic, N.N. Recent advances in the discovery of novel marine natural products and mycosporine-like amino acid UV-absorbing compounds. *Appl. Microbiol. Biotechnol.,* **2021**, *105*(19), 7053-7067. [http://dx.doi.org/10.1007/s00253-021-11467-9] [PMID: 34480237]

[12] Zhaohui, Z.; Xin, G.; Tashiro, Y.; Matsukawa, S.; Ogawa, H. Researches on the stability of porphyra-334 solution and its influence factors. *J. Ocean Univ. China,* **2004**, 3.

[13] Matsuyama, K.; Matsumoto, J.; Yamamoto, S.; Nagasaki, K.; Inoue, Y.; Nishijima, M.; Mori, T. pH-independent charge resonance mechanism for UV protective functions of shinorine and related mycosporine-like amino acids. *J. Phys. Chem. A,* **2015**, *119*(51), 12722-12729. [http://dx.doi.org/10.1021/acs.jpca.5b09988] [PMID: 26625701]

[14] Geraldes, V.; Pinto, E. Mycosporine-like amino acids (MAAs): Biology, chemistry and identification features. *Pharmaceuticals,* **2021**, *14*(1), 63. [http://dx.doi.org/10.3390/ph14010063] [PMID: 33466685]

CHAPTER 2

Distribution of MAAs

Abstract: Accumulation of mycosporine-like amino acids (MAAs) has been reported in a wide range of species in nature, including microalgae, macroalgae, cyanobacteria, phytoplankton, fungi, and some animals. This chapter describes the distribution of MAAs with a focus on macroalgae and cyanobacteria. MAAs biosynthesized by macroalgae have already been applied in cosmetic products, such as Helioguard 365 and Helinori. Macroalgae tend to accumulate multiple types of MAAs, and the types and accumulation levels are affected by changes in environmental factors. Regarding cyanobacteria, we focus on UV, salt, and osmotic stresses, temperature changes, and drought stress as environmental factors, and describe the species in which the accumulations of MAAs are induced by these stresses. UV-B irradiation is a common environmental factor that can induce the accumulation of MAAs in cyanobacteria, but induction by other abiotic stresses has been reported. These findings suggest that MAAs act as a multifunctional molecule that responds to a variety of environmental factors, not just as a UV absorber.

Keywords: Asterina-330, Cyanobacteria, Drought, Induction, Macroalgae, Mycosporine-2-glycine, Mycosporine-glycine, Osmotic stress, Porphyra-334, Pterin, Palythine, Red algae, Shinorine, Scytonemin, Temperature, UV-B.

INTRODUCTION

MAAs are widely distributed in nature. A wide variety of organisms, including microalgae, macroalgae, cyanobacteria, phytoplankton, and fungi, are known to biosynthesize MAAs. There have been no reports of intracellular accumulation of MAAs in bacteria (except cyanobacteria) or archaea, but a species of Actinomycetales, a Gram-positive bacterium, has been reported to contain trace amounts of shinorine depending on culture conditions [1]. MAAs have not been detected in higher plants, in which flavonoids act as UV-absorbing compounds. In animals, MAAs are sometimes detected, but this is thought to be due to uptake from other organisms through the food chain or symbiosis with microorganisms capable of biosynthesizing MAAs. However, it should be noted that the existence of homologs of cyanobacterial MAA synthetic genes has been reported in corals and sea anemones, and the possibility that these animals biosynthesize MAA can-

Hakuto Kageyama

not be ruled out [2]. In this chapter, the distribution of MAAs in marine macroalgae and cyanobacteria will be described.

MACROALGAE

Among macroalgae, red algae (rhodophytes) are known to accumulate MAAs. MAAs have also been detected in some green algae (chlorophytes) and brown algae (phaeophytes). According to a survey by Sun *et al.*, 572 species of macroalgae accumulating MAAs were reported in the 30 years from 1990 to 2019, of which 486 were red algae (Fig. **1**) [3]. In particular, MAAs were abundant in strains belonging to the orders Bangiales, Ceramiales, and Gracilariales [4].

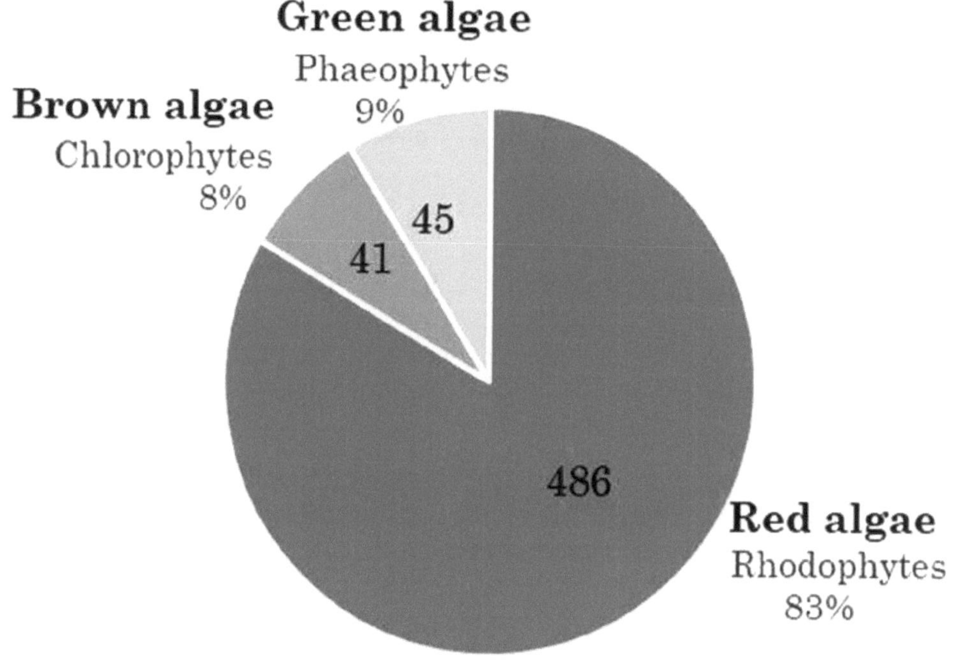

Fig. (1). Distribution of macroalgae that accumulate MAAs (Created based on the data in Sun Y *et al.* (2020) *Mar. Drugs*).

Seven types of MAAs are detected in red algae (mycosporine-glycine, porphyra-334, shinorine, palythine, palythene, palythinol, asterina-330), and the absorption wavelength range covered by these MAAs is wide (310–360 nm). Most red algae accumulate 4 to 5 types of MAA, which means that they can absorb a wide range of UV-A and UV-B [5]. In addition to these MAAs, usujirene has been reported to accumulate in species such as *Palmaria palmate*, *Gracilaria tenuifrons*, and *Porphyra yezoensis* [3]. In addition, in 2019, bostrychine A, B, C, D, E, and F were identified as new MAAs in *Bostrychia scorpioides* [6]. Red algae can adapt to the amount and types of MAAs accumulated and the regulation of biosynthesis of MAAs in response to fluctuating environmental factors, such as the degree of UV irradiation in the habitat, the concentration of nitrogen sources, salinity, and temperature. Helioguard 365 and Helinori, which are commercially available as cosmetic ingredients, contain MAAs extracted from the red alga *Porphyra umbilicalis*. Helioguard 365 is a formulation containing liposomal porphyra-334 and shinorine, and Helinori is a formulation containing porphyra-334, shinorine, and palythine. However, other MAAs extracted from red algae do not appear to be used in new product development so far. The main reason is that the content of MAAs in red algae collected from the sea is not sufficient. A maximum of 12 mg of MAAs content per gram of dry weight of red algae has been reported, but in most cases, less than half of that amount has been detected [5]. Exploration of algal strains with higher MAA contents or optimization of culture conditions that cause efficient biosynthesis of MAAs may lead to a solution. In addition, a method for efficient extraction and isolation of MAAs from red algae is required. From the viewpoint of industrial production, it is also considered effective to produce MAAs using microorganisms that are easy to culture and genetically transform, such as cyanobacteria.

For the identification of genes involved in MAA biosynthetic pathways and the characterization of the enzymatic reactions, analysis using cyanobacteria has preceded other organisms. However, in 2017, the cyanobacterial type MAA biosynthetic gene cluster was also reported in the genomic sequence of red algae including *Porphyra umbilicalis* and *Chondrus crispus* [7]. So far, there have been no further reports on the molecular regulatory mechanisms of these MAAs biosynthetic gene clusters in red algae, and future investigations are expected.

CYANOBACTERIA

In addition to algae, cyanobacteria are commonly used in the research of MAAs. Cyanobacteria are Gram-negative bacteria and are thought to have contributed greatly to the supply of oxygen and organic substances to the Earth by oxygen-evolving photosynthesis. Cyanobacteria are distributed throughout the waters and lands of the Earth. Also, being the primary producer of photosynthetic products in marine ecosystems, nitrogen-fixing cyanobacteria play a role as a nitrogen source in terrestrial ecosystems. Cyanobacteria also inhabit extreme environments such as deserts, hot springs, salt lakes, and polar regions [8]. In order to adapt to these extreme environments, cyanobacteria are thought to have acquired unique environmental adaptation strategies during their evolution. In particular, cyanobacteria are exposed to UV irradiation in sunlight when absorbing the solar energy required for photosynthesis. Therefore, it is considered that UV protection mechanisms have evolved at the molecular and cellular levels. One mechanism is the biosynthesis of UV-absorbing substances including MAAs. As mentioned in Chapter 1, MAAs can efficiently absorb mainly the UV-A and UV-B regions of the spectrum. The most frequently found MAA in cyanobacteria is shinorine, but MAAs with various molecular structures have been identified so far. MAAs exist in zwitterionic form due to the influence of amino acid residues present in the molecule, which impart high water solubility. Therefore, MAAs are generally localized in the cytoplasm. In addition, some species of cyanobacteria can biosynthesize a UV-abrorbing compound called scytonemin (Fig. **2**). The absorption maximum of scytonemin is around 370 nm, which corresponds to the UV-A region on the long wavelength side compared to MAAs, but it also has an absorption capacity in the UV-B and UV-C regions [8]. Unlike MAAs, scytonemin is a hydrophobic compound and is localized in the extracellular polysaccharide (EPS) matrix. Research on scytonemin has been intensive, and although it has not been completely elucidated, the biosynthetic pathway and its control mechanism have been roughly clarified. Although the details of scytonemin are not dealt with in this chapter, the outline of the biosynthetic pathway of scytonemin will be shown in an appendix to this book.

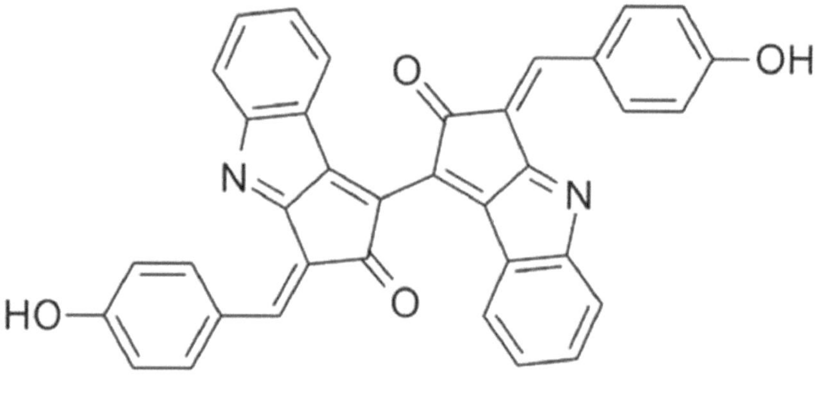

Fig. (2). Molecular structure of scytonemin (oxidized form).

Cyanobacteria in which MAA Accumulation is Induced by UV Irradiation

Intracellular accumulation of MAAs is induced in certain cyanobacteria when exposed to strong UV irradiation. For example, in 2008, it was reported that the filamentous cyanobacteria *Anabaena doliolum*, which can survive in sunny paddy fields during the summer, accumulated mycosporine-glycine, shinorine, and poriphyra-334, which were induced by UV-B irradiation [9]. Similarly, in the *Nodularia* strain of filamentous cyanobacteria (*Nodularia baltica*, *Nodularia harveyana*, *Nodularia spumigena*), it has been reported that the accumulation of shinorine and porphyra-334 was increased by UV-B irradiation. However, UV-A irradiation and photosynthetically active radiation (PAR) did not induce the accumulation of MAAs in the *Nodularia* strain [10]. In addition, a research group in Thailand reported similar research results in 2014. Firstly, it was clarified that UV-B irradiation induced the accumulation of shinorine and unidentified MAAs in the cyanobacteria *Gloeocapsa* sp. CU-2556 collected from a stone monument exposed to strong sunlight in the suburbs of Bangkok [11]. Secondly, *Arthrospira* sp. CU2556, collected from the same location, showed that the accumulation of mycosporine-glycine was not induced by UV-A but only by UV-B irradiation [12]. Thirdly, three types of MAAs, including palythine and asterina-330, were found in the cyanobacteria *Lyngbya* sp. CU2555 collected from the bark of a monkey pod plant (*Albizia saman*) in Bangkok, and the accumulation of these MAAs could also be induced by UV-B irradiation [13]. Besides, it has been reported that cyanobacteria, such as *Anabaena variabilis* ATCC29413 [14], *Chlorogloeopsis* PCC6912 [15], *Nostoc commune* [16], *Anabaena* sp [16], *Scytonema* sp [16], *Halothece* sp. PCC7418 [17], and *Nostoc flagelliforme* [18], accumulated MAAs in their cells. It has also been found that the accumulated levels of MAAs in these cyanobacterial species were increased by UV-B stimulation. Thus, in general, the bioproduction of MAAs is induced by UV-B stimulation in cyanobacteria. There was a report that UV-A irradiation induced three types of MAAs (mycosporine-taurine, dehydroxylusujirene, M-343) that accumulate in the freshwater cyanobacteria *Synechocystis* sp. PCC6803, but this is considered to be an exception [19]. Furthermore, given that the *Synechocystis* sp. PCC6803 strain does not carry any MAA biosynthetic genes, the result of MAA accumulation is questionable. As a UV-B receptor in cyanobacteria, an organic compound called pterin, which is composed of a pyrazine ring and a pyrimidine ring (Fig. **3**), has been proposed. In fact, treatment with a pterin inhibitor suppressed the induction of MAA accumulation by UV-B irradiation in the cyanobacterium *Chlorogloeopsis* PCC6912 [20]. Table **1** shows examples of cyanobacteria in which the accumulation of MAA is induced by UV-B irradiation.

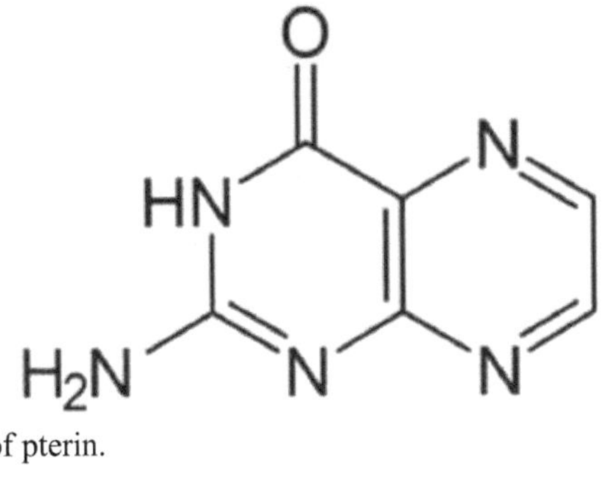

Fig. (3). Molecular structure of pterin.

Table 1. Cyanobacteria in which MAA accumulation is induced by UV-B irradiation.

Cyanobaterial Strain	Characteristics	Induced MAAs
Anabena doliolum [9]	A filamentous cyanobacterium that can grow under high sunshine conditions in summer.	Mycosporine-glycine Porphyra-334 Shinorine
Nodularia (*Nodularia baltica*, *Nodularia harveyana*, *Nodularia spumigena*) [10]	A nitrogen-fixing filamentous cyanobacterium. Inhabits brackish or salt water areas. Some strains produce a cyanotoxin called nodularin-R, which is harmful to the human body.	Shinorine Porphyra-334
Anabaena variabilis ATCC29413 [14]	A nitrogen-fixing filamentous cyanobacterium.	Shinorine
Chlorogloeopsis PCC6912 [15]	A filamentous cyanobacterium that can withstand salinity of up to 70% in seawater.	Shinorine Mycosporine-glycine
Nostoc commune [16]	A filamentous cyanobacterium that inhabits a variety of habitats and can survive for over 100 years after drying. It is thought that it can adapt to the land environment and arid areas under high UV irradiation conditions.	Shinorine (although other MAAs are accumulated, it is shinorine that was induced by UV-B irradiation in reference [16])
Anabaena sp [16].	A nitrogen-fixing filamentous cyanobacterium.	Shinorine
Scytonema sp [16].	A nitrogen-fixing cyanobacterial strain that grows into filaments. More than 100 species are known to exist, many of which are aquatic, but some grow on land. Some species are symbiotic with fungi and become lichens.	Shinorine
Gloeocapsa sp. CU-2556 [11]	A unicellular cyanobacterium isolated from a stone monument near Bangkok. It is exposed to high sunlight in the summer.	Shinorine M-307 (not characterized)
Arthrospira sp. CU2556 [12]	A filamentous cyanobacterium collected from a stone monument in Bangkok. The light intensity of the habitat is high.	Mycosporine-glycine
Lyngbya sp. CU2555 [13]	A filamentous cyanobacterium collected from the bark of a monkey pod plant in Bangkok.	Palythine Asterina-330 M-312 (not characterized)
Hassallia byssoidea [21]	A filamentous terrestrial cyanobacterium collected from a stone monument in India. It biosynthesizes hassallidin, a type of glycosylated lipoprotein that exhibits antifungal properties.	Mycosporine-alanine
Halothece sp. PCC7418 [17]	A halotolerant cyanobacterium isolated from the Dead Sea. It can grow under a wide range of NaCl concentrations (0.25–3.0 M).	Mycosporine-2-glycine

(Table 1) cont.....

Cyanobaterial Strain	Characteristics	Induced MAAs
Nostoc flagelliforme [18]	A drought-resistant terrestrial cyanobacterium. In some cases, it is edible.	Mycosporine-2-(4-deoxygadusol-ornithine)

Cyanobacteria in which MAA Accumulation is Induced by Salt and Osmotic Stress

When the salinity of the surrounding environment is high, cyanobacteria need to accumulate high concentrations of osmotic-compatible solutes in order to maintain the intracellular osmotic pressure. Such solutes are called compatible solutes. Well-known compatible solutes that cyanobacteria biosynthesize and accumulate in cells include sucrose, trehalose, glucosyl glycerol, and glycine betaine (Figs. **2**–**4**). Because MAAs are water-soluble, they may function not only as UV absorbers but also as compatible solutes. In the *Halothece* sp. PCC7418 strain*, which is a halotolerant cyanobacteria isolated from the Dead Sea in Israel, the intracellular accumulation of mycosporine-2-glycine increased significantly when the NaCl concentration in the medium was increased. In this strain, it was confirmed that the expression level of genes involved in the biosynthesis of mycosporine-2-glycine was increased by NaCl stress [17]. Given that the bioproduction of mycosporine-2-glycine was also induced by UV-B irradiation stress, it is possible that mycosporine-2-glycine is responsible for both UV protection and osmoregulation in this cyanobacterium. However, because glycine betaine is accumulated in high concentration as a compatible solute in this cyanobacteria strain, the contribution of mycosporin-2-glycine in maintaining osmotic pressure may not be so large. In fact, it was reported that glycine betaine was accumulated at a concentration about 1000 times higher than that of mycosporine-2-glycine [22]. However, when the biosynthetic genes of mycosporine-2-glycine were introduced into *Escherichia coli* so that mycosporine-2-glycine accumulated in the cells, it was possible to grow cells at a NaCl concentration higher than the growth limit NaCl concentration for wild-type strains [23]. Therefore, it is highly possible that mycosporine-2-glycine can function as an osmotic compatible solute in microbial cells such as *E. coli*. Interestingly, mycosporine-2-glycine has so far been detected only in halotolerant cyanobacterial strains [24].

* The *Halothece* sp. PCC7418 strain was originally isolated as *Aphanothece halophytica*. It can grow in a wide range of NaCl concentrations (0.25–3.0 M). It can also grow under basic conditions of pH 11. Under high salt concentrations, it is known that glycine betaine is biosynthesized and accumulated in cells at a high concentration by a three-step methylation reaction using glycine as a precursor.

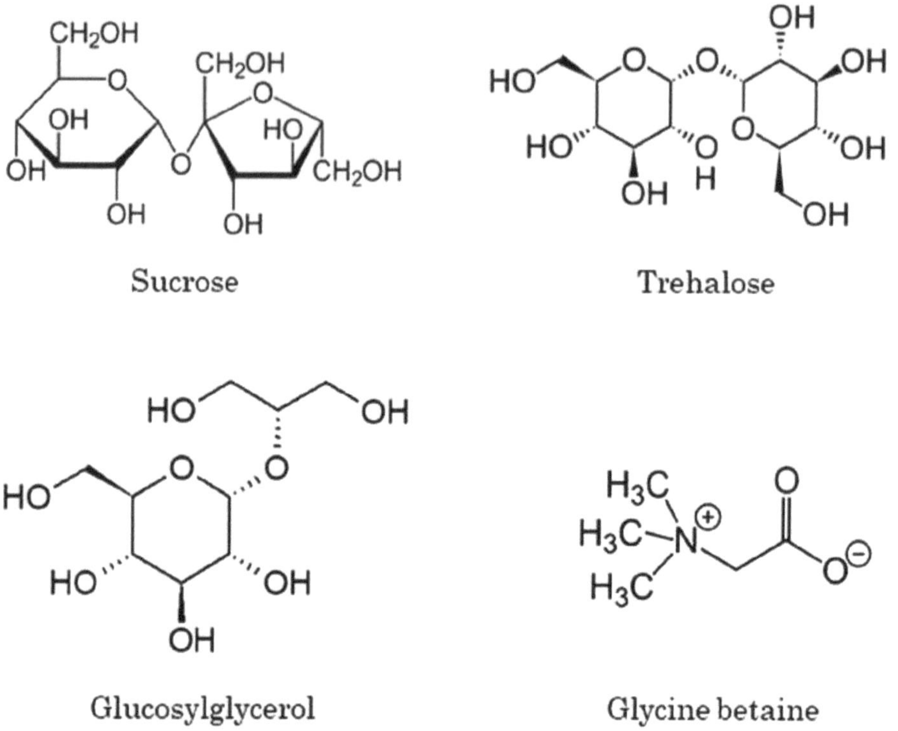

Sucrose

Trehalose

Glucosylglycerol

Glycine betaine

Fig. (4). Molecular structures of compatible solutes.

The filamentous cyanobacterium *Chlorogloeopsis* PCC6912 strain could withstand salinity of up to 70% in seawater. In this strain, in addition to UV-B stimulation, osmotic stress caused by NaCl and sucrose also induced the accumulation of mycosporine-glycine and shinorine [15]. However, similar to *Halothece* sp. PCC7418, the amount of these MAAs accumulated in NaCl-stressed *Chlorogloeopsis* PCC6912 cells was low compared to other major compatible solutes, such as sucrose and trehalose. The percentage of MAAs was only 5% of the total of these substances. Therefore, it is unlikely that MAAs contribute to osmoregulation in this strain.

Thus, it remains questionable whether MAAs function as compatible solutes in cyanobacterial cells. However, it is also a fact that most of the aquatic cyanobacteria capable of biosynthesizing MAAs live in a salt-containing environment such as the ocean, and it is rare that freshwater cyanobacteria accumulate MAAs. An example of a freshwater cyanobacterium that has been reported to accumulate MAAs is *Microcystis aeruginosa* [25]. *M. aeruginosa* is a unicellular cyanobacterium involved in the formation of harmful algal blooms in eutrophic waters. This cyanobacterial strain produces neurotoxins such as microcystin, including microcystin-LR (Fig. **5**), and cyanopeptolin. In this cyanobacterium, shinorine and porphyra-334 have been detected. Due to its habitat on the surface of the water, it may be synthesizing MAAs to protect cells from high exposure to solar irradiance, including UV. However, a report showed that UV irradiation treatment of *Microcystis aeruginosa* PCC7806 did not affect the accumulated amount of shinorine [26].

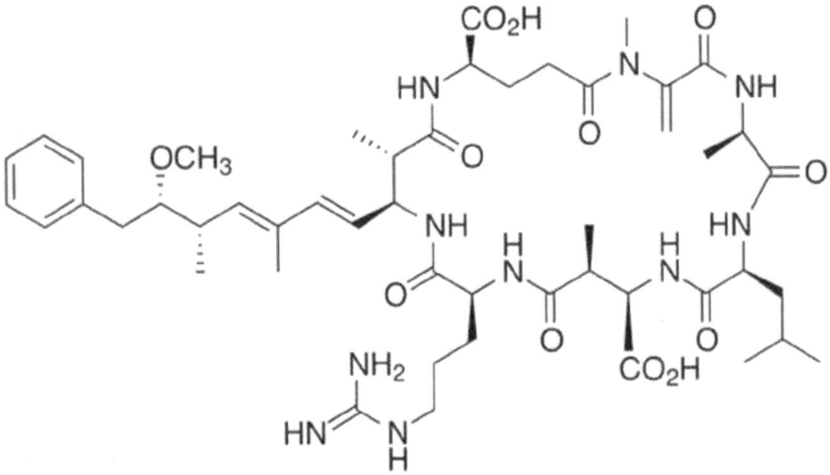

Fig. (5). Molecular structure of microcystin-LR.

Cyanobacteria in which MAA Accumulation is Induced by Temperature Changes

Temperature is one of the important fluctuation factors in the environment. However, it is considered that the involvement of temperature changes in the induction of MAA accumulation is not large in cyanobacteria. For example, the accumulation of mycosporine-glycine and shinorine in *Chlorogloeopsis* PCC6912 did not change even when cells were cultured at a temperature in the range of 10 to 50°C [15]. However, some cases have been reported in which the growth temperature affected the accumulation of MAAs. In *Anabaena variabilis* ATCC29413, shinorine accumulation induced by UV-B irradiation was markedly suppressed by treatment at 40°C (the control experiment was 20 ± 2°C) [14]. In the salt-tolerant cyanobacteria *Halothce* sp. PCC7418, a temperature shift from 30°C to 37°C had no effect, but a temperature shift to 23°C induced mycosporine-2-glycine accumulation [27]. Other than cyanobacteria, heat stress has been reported to increase MAA accumulation in the corals *Lobophytum compactum* and *Sunularia flexibilis* [28].

Cyanobacteria of the Genus *Nostoc* that are Resistant to Drought

Cyanobacteria of the genus *Nostoc* (Fig. **6**) are filamentous cyanobacteria that form colonies covered with gelatinous sheaths. They are nitrogen-fixing cyanobacteria that differentiate into cells called heterocysts and take in nitrogen from the air. These cells are edible. For example, in Japan, *Nostoc commune* is cooked in vinegar and eaten. *Nostoc* strains are exposed to environmental stresses such as strong sunlight, high temperature, and dryness in their habitat. On land areas, they are found on the surface of soil, moist rocks, and concrete. In water areas, they inhabit the bottom of lakes and springs. Although rarely seen in the ocean, they show some salt tolerance. They are extremely resistant to dryness, and even if they have been kept dry for more than 100 years, they will restart their vital activity when immersed in culture medium. It is well known that *Nostoc* accumulates MAAs. There are many reports of MAA derivatized by the addition of molecules such as sugar. Table **2** lists the MAAs detected so far in *Nostoc commune*. There have been many reports on derivatized MAAs since the beginning of the 2010s. Recently, it has been revealed that there is a unique combination of MAA synthetic genes in drought-tolerant cyanobacteria, including *Nostoc commune*. This gene combination has been shown to be involved in the biosynthesis of MAAs with complex structures [29], which will be described in Chapter 3.

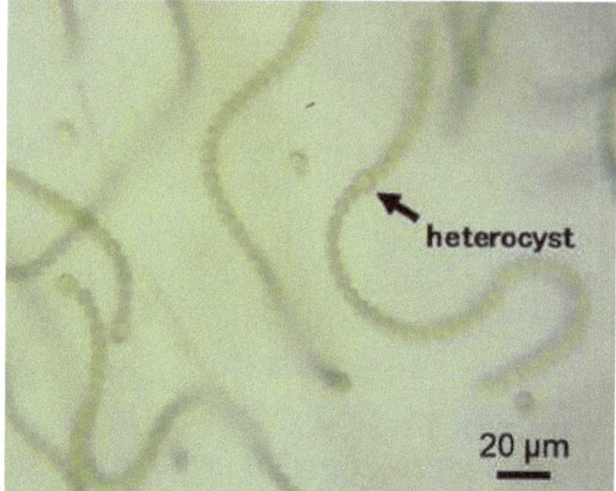

Fig. (6). Microscopic image of *Nostoc commune*.

CONCLUDING REMARKS

This chapter described the distribution of MAAs with a focus on macroalgae and cyanobacteria. Although MAAs biosynthesized by macroalgae have already been commercialized, in the future, more efficient preparation conditions for MAAs might be established due to the different compositions and amounts of MAA accumulation depending on the strain and environmental factors in the habitat. In cyanobacteria, although it has been reported that various strains can biosynthesize

MAAs, many strains cannot. In cyanobacteria, biosynthesis of MAAs is induced by UV-B irradiation and is affected by salt and environmental factors such as osmotic stress and temperature changes. The MAA synthetic pathway reactions will be described in detail in Chapter 3, but many regulatory mechanisms remain to be elucidated.

Table 2. MAAs detected in *Nostoc commune*.

Detected MAAs (Chronological Order)
Shinorine [16, 30]
450-Da MAA [31] (Hexose-bound palythine-threonine)
612-Da MAA [31] (Two hexose-bound palythine-threonine)
508-Da MAA [31, 32] (Hexose-bound porphyra-334)
Mycosporine-GABA [32]
464-Da MAA [32] (Pentose-bound shinorine)
478-Da MAA [32] (7-*O*-(β-arabinopyranosyl)-porphyra-334)
880-Da MAA [32] ({Mycosporine-ornithine:4-deoxygadusol ornithine} -β-xylopyranosyl-β-galactopyranoside)
1050-Da MAA [32] (Mycosporine-2-(4-deoxygadusol-ornithine)-β-xylopyranosyl-β-galactopyranoside)
Porphyra-334 [32-34] (Also detected in *Nostoc sphericum* and *Nostoc verrucosum*)
756-Da MAA [35] (Mycosporine-2-(4-deoxygadusol-ornithine), Nostoc-756)

LIST OF ABBREVIATIONS

MAA mycosporine-like amino acid

UV ultraviolet

CONSENT FOR PUBLICATION

Not applicable.

CONFLICT OF INTEREST

The author declares no conflict of interest, financial or otherwise.

ACKNOWLEDGEMENTS

Declared none.

REFERENCES

[1] Miyamoto, K.T.; Komatsu, M.; Ikeda, H. Discovery of gene cluster for mycosporine-like amino acid biosynthesis from *Actinomycetales* microorganisms and production of a novel mycosporine-like amino acid by heterologous expression. *Appl. Environ. Microbiol.*, **2014**, *80*(16), 5028-5036. [http://dx.doi.org/10.1128/AEM.00727-14] [PMID: 24907338]

[2] Shinzato, C.; Shoguchi, E.; Kawashima, T.; Hamada, M.; Hisata, K.; Tanaka, M.; Fujie, M.; Fujiwara, M.; Koyanagi, R.; Ikuta, T.; Fujiyama, A.; Miller, D.J.; Satoh, N. Using the *Acropora digitifera* genome to understand coral responses to environmental change. *Nature,* **2011**, *476*(7360), 320-323. [http://dx.doi.org/10.1038/nature10249] [PMID: 21785439]

[3] Sun, Y.; Zhang, N.; Zhou, J.; Dong, S.; Zhang, X.; Guo, L.; Guo, G. Distribution, contents, and types of mycosporine-like amino acids (MAAs) in marine macroalgae and a database for MAAs based on these characteristics. *Mar. Drugs,* **2020**, *18*(1), 43. [http://dx.doi.org/10.3390/md18010043] [PMID: 31936139]

[4] Sun, Y.; Han, X.; Hu, Z.; Cheng, T.; Tang, Q.; Wang, H.; Deng, X.; Han, X. Extraction, isolation and characterization of mycosporine-like amino acids from four species of red macroalgae. *Mar. Drugs,* **2021**, *19*(11), 615. [http://dx.doi.org/10.3390/md19110615] [PMID: 34822486]

[5] Navarro, N.; Figueroa, F.L.; Korbee, N.; Bonomi, J.; Gomez, F.A.; de la Coba, F. Mycosporine-like amino acids from red algae to develop natural UV sunscreens. In: *Sunscreens: Source, Formulations, Efficacy and Recommendations*; Rastogi, R.P., Ed.; Nova Science Publishers, **2018**.

[6] Orfanoudaki, M.; Hartmann, A.; Alilou, M.; Gelbrich, T.; Planchenault, P.; Derbré, S.; Schinkovitz, A.; Richomme, P.; Hensel, A.; Ganzera, M. Absolute configuration of mycosporine-like amino acids, their wound healing properties and *in vitro* anti-aging effects. *Mar. Drugs,* **2019**, *18*(1), 35. [http://dx.doi.org/10.3390/md18010035] [PMID: 31906052]

[7] Brawley, S.H.; Blouin, N.A.; Ficko-Blean, E.; Wheeler, G.L.; Lohr, M.; Goodson, H.V.; Jenkins, J.W.; Blaby-Haas, C.E.; Helliwell, K.E.; Chan, C.X.; Marriage, T.N.; Bhattacharya, D.; Klein, A.S.; Badis, Y.; Brodie, J.; Cao, Y.; Collén, J.; Dittami, S.M.; Gachon, C.M.M.; Green, B.R.; Karpowicz, S.J.; Kim, J.W.; Kudahl, U.J.; Lin, S.; Michel, G.; Mittag, M.; Olson, B.J.S.C.; Pangilinan, J.L.; Peng, Y.; Qiu, H.; Shu, S.; Singer, J.T.; Smith, A.G.; Sprecher, B.N.; Wagner, V.; Wang, W.; Wang, Z.Y.; Yan, J.; Yarish, C.; Zäuner-Riek, S.; Zhuang, Y.; Zou, Y.; Lindquist, E.A.; Grimwood, J.; Barry, K.W.; Rokhsar, D.S.; Schmutz, J.; Stiller, J.W.; Grossman, A.R.; Prochnik, S.E. Insights into the red algae and eukaryotic evolution from the genome of *Porphyra umbilicalis* (Bangiophyceae, Rhodophyta). *Proc. Natl. Acad. Sci. USA,* **2017**, *114*(31), E6361-E6370. [http://dx.doi.org/10.1073/pnas.1703088114] [PMID: 28716924]

[8] Kageyama, H.; Waditee-Sirisattha, R. Cyanobacterial UV sunscreen: Biosynthesis, regulation, and application. In: *Sunscreens: Source, Formulations, Efficacy and Recommendations*; Rastogi, R.P., Ed.; NOVA Science Publishers, **2018**; pp. 1-28.

[9] Singh, S.P.; Sinha, R.P.; Klisch, M.; Häder, D.P. Mycosporine-like amino acids (MAAs) profile of a rice-field cyanobacterium *Anabaena doliolum* as influenced by PAR and UVR. *Planta,* **2008**, *229*(1), 225-233. [http://dx.doi.org/10.1007/s00425-008-0822-1] [PMID: 18830707]

[10] Sinha, R.P.; Ambasht, N.K.; Sinha, J.P.; Klisch, M.; Häder, D.P. UV-B-induced synthesis of mycosporine-like amino acids in three strains of Nodularia (cyanobacteria). *J. Photochem. Photobiol. B,* **2003**, *71*(1-3), 51-58. [http://dx.doi.org/10.1016/j.jphotobiol.2003.07.003] [PMID: 14705639]

[11] Rastogi, R.P.; Incharoensakdi, A. UV radiation-induced biosynthesis, stability and antioxidant activity of mycosporine-like amino acids (MAAs) in a unicellular cyanobacterium *Gloeocapsa* sp. CU2556. *J. Photochem. Photobiol. B,* **2014**, *130*, 287-292. [http://dx.doi.org/10.1016/j.jphotobiol.2013.12.001] [PMID: 24374576]

[12] Rastogi, R.P.; Incharoensakdi, A. Analysis of UV-absorbing photoprotectant mycosporine-like amino acid (MAA) in the cyanobacterium *Arthrospira* sp. CU2556. *Photochem. Photobiol. Sci.,* **2014**, *13*(7), 1016-1024. [http://dx.doi.org/10.1039/c4pp00013g] [PMID: 24769912]

[13] Rastogi, R.P.; Incharoensakdi, A. Characterization of UV-screening compounds, mycosporine-like amino acids, and scytonemin in the cyanobacterium *Lyngbya* sp. CU2555. *FEMS Microbiol. Ecol.,* **2014**, *87*(1), 244-256. [http://dx.doi.org/10.1111/1574-6941.12220] [PMID: 24111939]

[14] Singh, S.P.; Klisch, M.; Sinha, R.P.; Häder, D.P. Effects of abiotic stressors on synthesis of the mycosporine-like amino acid shinorine in the Cyanobacterium *Anabaena variabilis* PCC 7937. *Photochem. Photobiol.,* **2008**, *84*(6), 1500-1505. [http://dx.doi.org/10.1111/j.1751-1097.2008.00376.x] [PMID: 18557824]

[15] Portwich, A.; Garcia-Pichel, F. Ultraviolet and osmotic stresses induce and regulate the synthesis of mycosporines in the cyanobacterium *Chlorogloeopsis* PCC 6912. *Arch. Microbiol.,* **1999**, *172*(4), 187-192. [http://dx.doi.org/10.1007/s002030050759] [PMID: 10525734]

[16] Sinha, R.P.; Klisch, M.; Walter Helbling, E.; Häder, D.P. Induction of mycosporine-like amino acids (MAAs) in cyanobacteria by solar ultraviolet-B radiation. *J. Photochem. Photobiol. B,* **2001**, *60*(2-3), 129-135. [http://dx.doi.org/10.1016/S1011-1344(01)00137-3] [PMID: 11470569]

[17] Waditee-Sirisattha, R.; Kageyama, H.; Sopun, W.; Tanaka, Y.; Takabe, T. Identification and upregulation of biosynthetic genes required for accumulation of Mycosporine-2-glycine under salt stress conditions in the halotolerant cyanobacterium *Aphanothece halophytica. Appl. Environ. Microbiol.,* **2014**, *80*(5), 1763-1769. [http://dx.doi.org/10.1128/AEM.03729-13] [PMID: 24375141]

[18] Shang, J.L.; Zhang, Z.C.; Yin, X.Y.; Chen, M.; Hao, F.H.; Wang, K.; Feng, J.L.; Xu, H.F.; Yin, Y.C.; Tang, H.R.; Qiu, B.S. UV-B induced biosynthesis of a novel sunscreen compound in solar radiation and desiccation tolerant cyanobacteria. *Environ. Microbiol.,* **2018**, *20*(1), 200-213. [http://dx.doi.org/10.1111/1462-2920.13972] [PMID: 29076601]

[19] Zhang, L.; Li, L.; Wu, Q. Protective effects of mycosporine-like amino acids of *Synechocystis* sp. PCC 6803 and their partial characterization. *J. Photochem. Photobiol. B,* **2007**, *86*(3), 240-245. [http://dx.doi.org/10.1016/j.jphotobiol.2006.10.006] [PMID: 17182253]

[20] Portwich, A.; Garcia-Pichel, F. A novel prokaryotic UVB photoreceptor in the cyanobacterium *Chlorogloeopsis* PCC 6912. *Photochem. Photobiol.,* **2000**, *71*(4), 493-498. [http://dx.doi.org/10.1562/0031-8655(2000)071<0493:ANPUPI>2.0.CO;2] [PMID: 10824604]

[21] Saha, S.; Sen, A.; Mandal, S.; Adhikary, S.P.; Rath, J. Mycosporine-alanine, an oxo-mycosporine, protect *Hassallia byssoidea* from high UV and solar irradiation on the stone monument of Konark. *J. Photochem. Photobiol. B,* **2021**, *224*, 112302. [http://dx.doi.org/10.1016/j.jphotobiol.2021.112302] [PMID: 34537544]

[22] Waditee-Sirisattha, R.; Kageyama, H.; Fukaya, M.; Rai, V.; Takabe, T. Nitrate and amino acid availability affects glycine betaine and mycosporine-2-glycine in response to changes of salinity in a halotolerant cyanobacterium *Aphanothece halophytica. FEMS Microbiol. Lett.,* **2015**, *362*(23), fnv198. [http://dx.doi.org/10.1093/femsle/fnv198] [PMID: 26474598]

[23] Patipong, T.; Hibino, T.; Waditee-Sirisattha, R.; Kageyama, H. Efficient bioproduction of mycosporine-2-glycine, which functions as potential osmoprotectant, using *Escherichia coli* cells. *Nat.*

Prod. Commun., **2017**, *12*(10), 1934578X1701201.
[http://dx.doi.org/10.1177/1934578X1701201017]

[24] Kedar, L.; Kashman, Y.; Oren, A. Mycosporine-2-glycine is the major mycosporine-like amino acid in a unicellular cyanobacterium (*Euhalothece* sp.) isolated from a gypsum crust in a hypersaline saltern pond. *FEMS Microbiol. Lett.*, **2002**, *208*(2), 233-237.
[http://dx.doi.org/10.1111/j.1574-6968.2002.tb11087.x] [PMID: 11959442]

[25] Volkmann, M.; Gorbushina, A.A.; Kedar, L.; Oren, A. Structure of euhalothece-362, a novel red-shifted mycosporine-like amino acid, from a halophilic cyanobacterium (Euhalothece sp.). *FEMS Microbiol. Lett.*, **2006**, *258*(1), 50-54.
[http://dx.doi.org/10.1111/j.1574-6968.2006.00203.x] [PMID: 16630254]

[26] Liu, Z.; Hader, D.P.; Sommaruga, R. Occurrence of mycosporine-like amino acids (MAAs) in the bloom-forming cyanobacterium *Microcystis aeruginosa. J. Plankton Res.*, **2004**, *26*(8), 963-966.
[http://dx.doi.org/10.1093/plankt/fbh083]

[27] Hu, C.; Völler, G.; Süßmuth, R.; Dittmann, E.; Kehr, J.C. Functional assessment of mycosporine-like amino acids in *M icrocystis aeruginosa* strain PCC 7806. *Environ. Microbiol.*, **2015**, *17*(5), 1548-1559.
[http://dx.doi.org/10.1111/1462-2920.12577] [PMID: 25059440]

[28] Patipong, T.; Hibino, T.; Waditee-Sirisattha, R.; Kageyama, H. Induction of antioxidative activity and antioxidant molecules in the halotolerant cyanobacterium *Halothece* sp. PCC7418 by temperature shift. *Nat. Prod. Commun.*, **2019**, *14*(7), 1934578X1986568.
[http://dx.doi.org/10.1177/1934578X19865680]

[29] Shick, J.M.; Dunlap, W.C. Mycosporine-like amino acids and related Gadusols: Biosynthesis, acumulation, and UV-protective functions in aquatic organisms. *Annu. Rev. Physiol.*, **2002**, *64*(1), 223-262.
[http://dx.doi.org/10.1146/annurev.physiol.64.081501.155802] [PMID: 11826269]

[30] Zhang, Z.C.; Wang, K.; Hao, F.H.; Shang, J.L.; Tang, H.R.; Qiu, B.S. New types of ATP -grasp ligase are associated with the novel pathway for complicated mycosporine-like amino acid production in desiccation-tolerant cyanobacteria. *Environ. Microbiol.*, **2021**, *23*(11), 6420-6432.
[http://dx.doi.org/10.1111/1462-2920.15732] [PMID: 34459073]

[31] Sinha, R.P.; Ambasht, N.K.; Sinha, J.P.; Häder, D.P. Wavelength-dependent induction of a mycosporine-like amino acid in a rice-field cyanobacterium, *Nostoc commune*: Role of inhibitors and salt stress. *Photochem. Photobiol. Sci.*, **2003**, *2*(2), 171-176.
[http://dx.doi.org/10.1039/b204167g] [PMID: 12664980]

[32] Nazifi, E.; Wada, N.; Yamaba, M.; Asano, T.; Nishiuchi, T.; Matsugo, S.; Sakamoto, T. Glycosylated porphyra-334 and palythine-threonine from the terrestrial cyanobacterium *Nostoc commune. Mar. Drugs,* **2013**, *11*(9), 3124-3154.
[http://dx.doi.org/10.3390/md11093124] [PMID: 24065157]

[33] Nazifi, E.; Wada, N.; Asano, T.; Nishiuchi, T.; Iwamuro, Y.; Chinaka, S.; Matsugo, S.; Sakamoto, T. Characterization of the chemical diversity of glycosylated mycosporine-like amino acids in the terrestrial cyanobacterium *Nostoc commune. J. Photochem. Photobiol. B,* **2015**, *142*, 154-168.
[http://dx.doi.org/10.1016/j.jphotobiol.2014.12.008] [PMID: 25543549]

[34] Ishihara, K.; Watanabe, R.; Uchida, H.; Suzuki, T.; Yamashita, M.; Takenaka, H.; Nazifi, E.; Matsugo, S.; Yamaba, M.; Sakamoto, T. Novel glycosylated mycosporine-like amino acid, 13- O -(- -galactosyl)-porphyra-334, from the edible cyanobacterium Nostoc sphaericum -protective activity on human keratinocytes from UV light. *J. Photochem. Photobiol. B,* **2017**, *172*, 102-108.
[http://dx.doi.org/10.1016/j.jphotobiol.2017.05.019] [PMID: 28544967]

[35] Inoue-Sakamoto, K.; Nazifi, E.; Tsuji, C.; Asano, T.; Nishiuchi, T.; Matsugo, S.; Ishihara, K.; Kanesaki, Y.; Yoshikawa, H.; Sakamoto, T. Characterization of mycosporine-like amino acids in the cyanobacterium <i>Nostoc verrucosum</i>. *J. Gen. Appl. Microbiol.*, **2018**, *64*(5), 203-211.

[http://dx.doi.org/10.2323/jgam.2017.12.003] [PMID: 29709901]

[36] Sakamoto, T.; Hashimoto, A.; Yamaba, M.; Wada, N.; Yoshida, T.; Inoue-Sakamoto, K.; Nishiuchi, T.; Matsugo, S. Four chemotypes of the terrestrial cyanobacterium *Nostoc commune* characterized by differences in the mycosporine-like amino acids. *Phycol. Res.,* **2019**, *67*(1), 3-11. [http://dx.doi.org/10.1111/pre.12333]

CHAPTER 3

Biosynthetic Pathways of MAAs and their Regulatory Mechanisms

Abstract: The biosynthetic mechanism of mycosporine-like amino acids (MAAs) has been roughly elucidated. In 2010, the genes responsible for MAA biosynthesis were identified in cyanobacteria. In this chapter, first, we will describe the reaction mechanisms responsible for the biosynthetic pathways of MAAs, mainly based on results from cyanobacteria. Next, as a regulatory mechanism for MAA biosynthesis, the response patterns of MAA accumulation in response to abiotic stresses, such as UV irradiation, salt, and osmotic pressure, will be explained. There are many points to be clarified regarding the detailed regulatory mechanisms, and further analyses are awaited in the future. Because MAAs have useful activities in addition to UV absorption, they are substances that are expected to be used in cosmetics and pharmaceuticals. This chapter also includes discussions from the perspective of future industrial production.

Keywords: ATP-grasp enzyme, Biosynthetic pathways of MAAs, D-Ala-D-Ala ligase, Gene resources, Nonribosomal peptide synthetase (NRPS), Localization, Osmotic stress, Pentose phosphate pathway, Shikimate pathway, Regulatory mechanism, Substrate specificity, Salt stress, Temperature, UV irradiation, 4-deoxygadusol, 3-dehydroquinate synthase (DHQS), *O*–methyltransferase (*O*–MT), 2-*epi*-5-*epi*-valiolone synthase (EVS).

INTRODCUTION

Biosynthetic Pathways of MAAs

The biosynthetic pathways of MAAs have been intensively analyzed using cyanobacteria. It is known that the precursor compounds of MAAs are produced from metabolic intermediates of primary metabolic pathways. Mono- or di-substituted MAAs are produced based on the precursor compound through several enzymatic reactions.

Hakuto Kageyama

Biosynthetic Pathway of 4-deoxygadusol, A Precursor Compound of MAAs

In cyanobacteria, the precursor compound of MAAs is 4-deoxygadusol (4-DG) (Fig. **1**). 4-DG is thought to be produced from metabolic intermediates of theprimary metabolic pathways, the shikimate pathway, or the pentose phosphate pathway.

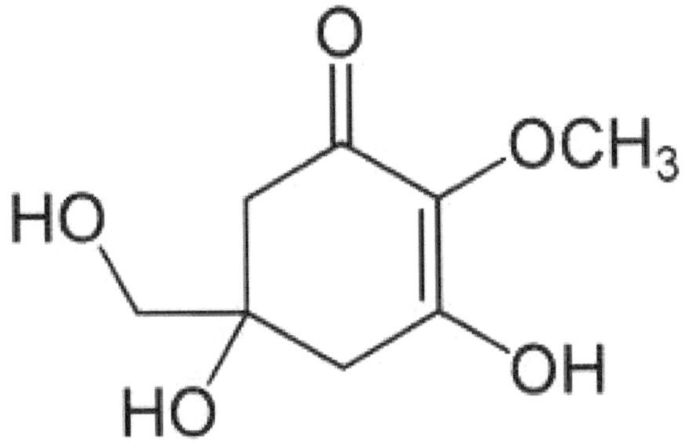

Fig. (1). Molecular structure of 4-deoxygadusol (4-DG).

The shikimate pathway (Fig. **2**) is a biosynthetic reaction pathway for the aromatic amino acids tyrosine, phenylalanine, and tryptophan, which has not been found in animals but is present in most microorganisms and plants. The pentose phosphate pathway (Fig. **3**) is a pathway involved in the production of various pentoses. In this pathway, glucose-6-phosphate (G6P), which is an intermediate of glycolysis, is converted into glyceraldehyde-3-phosphate (G3P), which is also an intermediate of glycolysis. In the pentose phosphate pathway, one molecule of G6P produces one molecule of CO_2 and two molecules of NADPH. Therefore, it is a source of NADPH*.

* NADPH is a reduced form of nicotinamide adenine dinucleotide phosphate (NADP), which is used as an electron carrier in photosynthetic pathways and glycolysis.

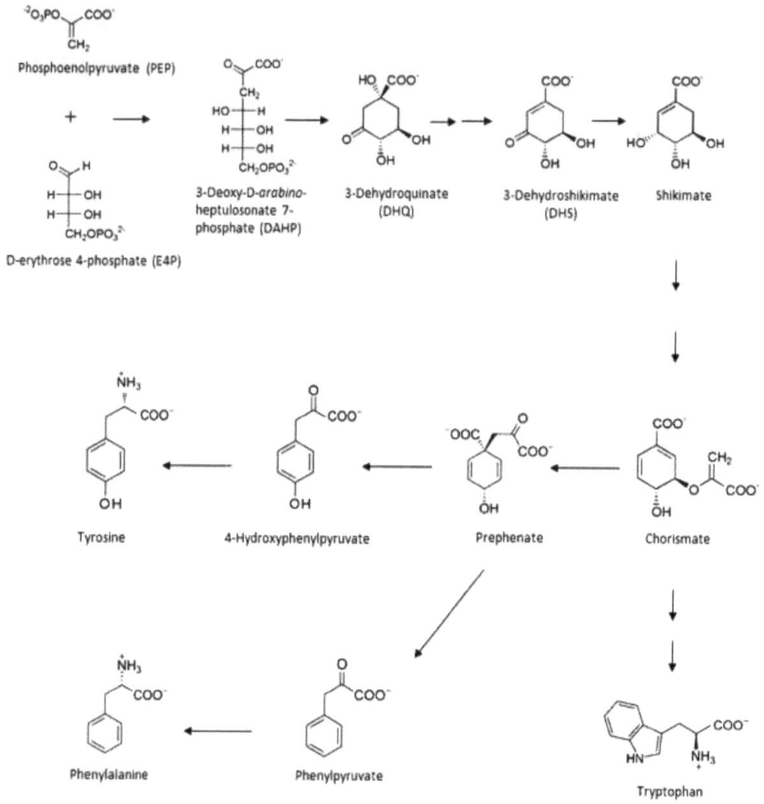

Fig. (2). The shikimate pathway.

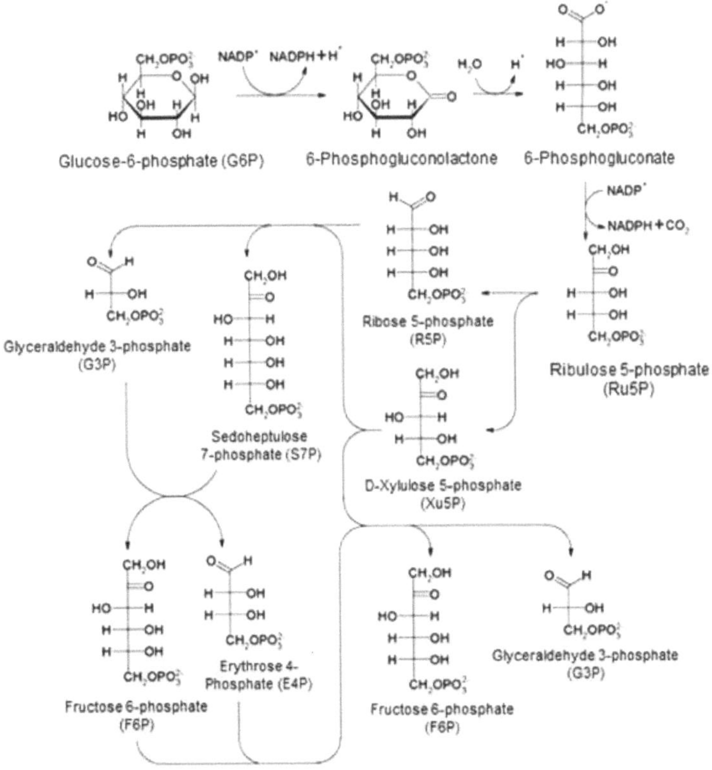

Fig. (3). The pentose phosphate pathway.

Biosynthesis of 4-DG from the Shikimate Pathway

In the shikimate pathway, 3-deoxy-D-arabinohepturonic acid 7-phosphate (DAHP) is cyclized and converted into 3-dehydroquinate (DHQ) by 3-dehydroquinate synthase (DHQS) (Fig. 4). Furthermore, although the reaction mechanism has not been clarified, it is thought that demethyl-4-deoxygadusol (DDG) is produced by this catalytic action of DHQS. Subsequently, 4-DG is produced by the catalytic action of *O*–methyltransferase (*O*–MT). *O*–MT converts the hydroxyl group (–OH) at the C2-position into a methoxy group (–OCH_3).

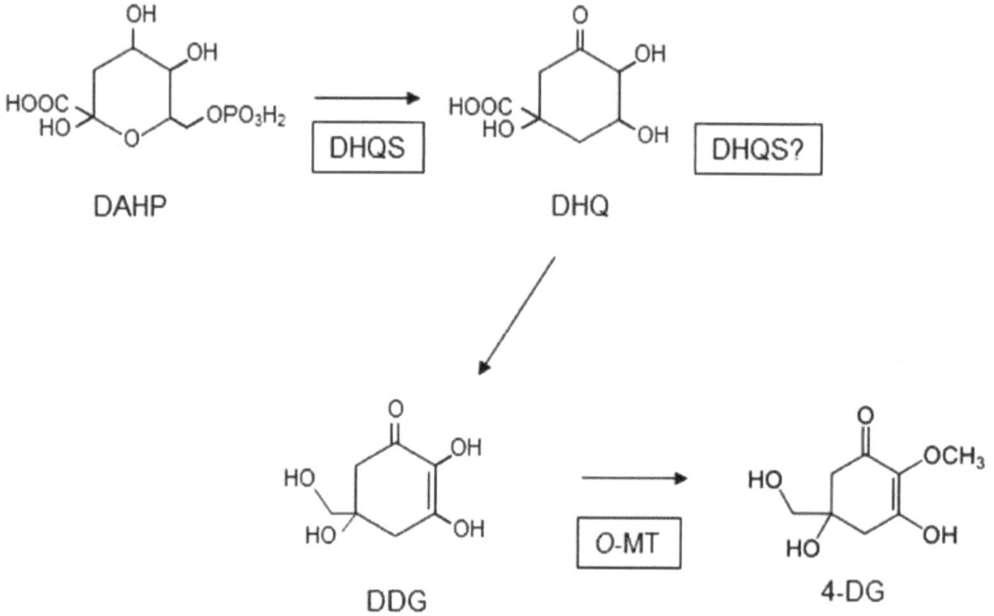

Fig. (4). The 4-DG biosynthetic pathway following the shikimate pathway.

The following experimental results showed that the shikimate pathway is involved in 4-DG biosynthesis.

i. Incorporation experiments using radioisotopes in the genus *Trichothecium roseum* showed that DHQ is a precursor of the cyclohexenone ring structure of mycosporine [1].
ii. Pyruvate, a precursor of phosphoenolpyruvate (PEP), was labeled with a ^{14}C isotope and incorporated into the cyanobacterium *Chlorogloeopsis* PCC6912 strain. As a result, the ^{14}C isotope was detected in the cyclohexenone structure of MAAs [2]. Phosphoenolpyruvate is an intermediate in the shikimate pathway.
iii. When tyrosine was added to the culture of the cyanobacteria *Chlorogloeopsis* PCC6912 strain, the biosynthesis of MAAs was inhibited. Excessive amounts of tyrosine are known to inhibit the shikimate pathway [2].
iv. In the coral *Stylophora pistillata*, the synthesis of MAAs was inhibited by treatment with glyphosate, which is a shikimate pathway-specific inhibitor [3].

Discovery of DHQS and O-MT by Genome Mining

The entire genome sequence of many species has been clarified, and the search for genes with specific functions by comparing their genome information is called "genome mining". The genome sequences of many strains of cyanobacteria have also been published. By genome mining using these genes, genes encoding DHQS and *O*–MT involved in 4-DG biosynthesis were searched by Singh *et al.* in 2010, and candidate genes were proposed. They found that *Anabaena variabilis* ATCC29413 (*Anabaena variabilis* PCC7937), which can synthesize MAAs, possessed an adjacent set of the *DHQS* gene (Ava_3858) and *O*–MT gene (Ava_3857), but the cyanobacterial strains, including *Anabaena* sp PCC7120, *Synechococcus* sp. PCC6301, and *Synechocystis* sp. PCC6803, which cannot synthesize MAAs, did not have this gene set. These findings suggested that these genes are involved in 4-DG biosynthesis [4].

Biosynthesis of 4-DG from the Pentose Phosphate Pathway

In 2010, the same year that the DHQS and *O*–MT genes were identified, which are responsible for 4-DG biosynthesis following the shikimate pathway, a 4-DG biosynthesis pathway following the pentose phosphate pathway was reported. In this pathway, a series of reactions begin with sedoheptulose-7-phophate (S7P), which is an intermediate of the pentose phosphate pathway (Fig. **5**). First, 2-*epi*-5-*epi*-valiolone synthase (EVS) converts S7P to DDG *via* 2-*epi*-5-*epi*-valiolone (EV) using S7P as a substrate. *O*–MT then catalyzes the 4-DG formation reaction. These reactions were confirmed *in vitro* using enzymes derived from the cyanobacterium *Nostoc punctiforme* ATCC29133 (EVS: NpR5600, *O*–MT: NpR5599) that were expressed in *E. coli* and then purified [5]. At that time, 4-DG was not obtained by reaction with DAHP, which is an intermediate of the shikimate pathway, as a substrate. It should be noted that these results do not deny that 4-DG is produced following the shikimate pathway. In cyanobacterial cells, the 4-DG production reaction following the shikimate pathway may proceed when the conditions necessary for the reaction are met.

In cyanobacterial cells, 4-DG may be biosynthesized following both the shikimate and pentose phosphate pathways. Both DHQS and EVS are enzymes classified into the superfamily of sugar phosphate cyclase, and their amino acid sequences are relatively similar. Therefore, in cyanobacterial cells, DHQS/EVS may be able to recognize both DHAP (an intermediate of the shikimate pathway) and S7P (an intermediate of the pentose phosphate pathway) as substrates.

It was reported that in the cyanobacterium *A. variabilis* ATCC29413 strain, the biosynthetic ability of MAAs remained even when the *Ava_3858* gene encoding DHQS/EVS was deleted. When this gene-deficient strain was treated with glyphosate or phenylalanine, which is an inhibitor of the shikimate pathway, MAAs could not be produced. Thus, it was strongly suggested that MAAs were biosynthesized from the shikimate pathway in this strain [6]. It was thought that an enzyme alternative to Ava_3858 catalyzed the relevant reaction in MAA biosynthesis in this gene-deficient cyanobacterial strain, but there have been no reports of genes encoding that enzyme. However, when the *Ava_3857* gene encoding *O*–MT was deleted, the biosynthetic ability of MAAs was lost [7]. Thus, *O*-MT was considered to be an essential factor in the MAA biosynthetic pathway following both the shikimate and pentose phosphate pathways.

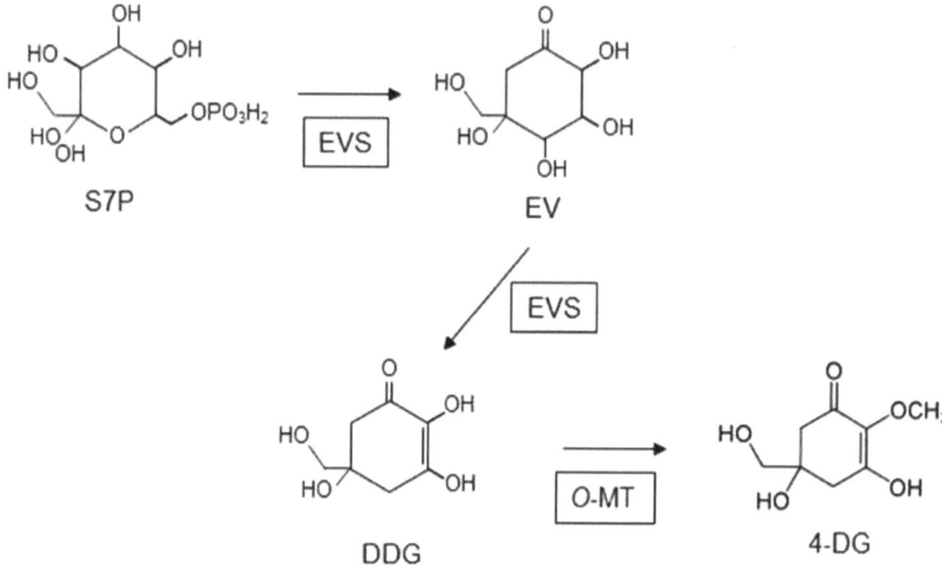

Fig. (5). The 4-DG biosynthetic pathway following the pentose phosphate pathway.

Biosynthetic Pathway of MAAs Utilizing 4-DG as a Precursor

Substitution of the hydroxyl group at the C3 position of 4-DG with glycine produces mycosporine-glycine. This reaction is catalyzed by an enzyme homologous to the superfamily of the ATP-grasp enzyme, which corresponds to the *Ava_3856* gene in *A. variabilis* ATCC29413. The *Ava_3856* gene is located adjacent to the *Ava_3857* gene, which encodes *O*–MT. As shown in (Fig. **6**), the Ava_3856 protein activates 4-DG by phosphorylation with the phosphate group of ATP. The Ava_3856 protein then adds glycine to the acivated 4-DG to generate a monosubstituted MAA, mycosporine-glycine, by a 1,4-addition reaction [5]*.

* A double bond is the site of an addition reaction. When an atom or molecule bonds to each atom that forms a double bond, it is called 1,2-addition (direct addition). In a conjugated diene, not only direct addition but also 1,4-addition (conjugate addition) occurs in which an addition reaction occurs for each terminal atom constituting the conjugate diene.

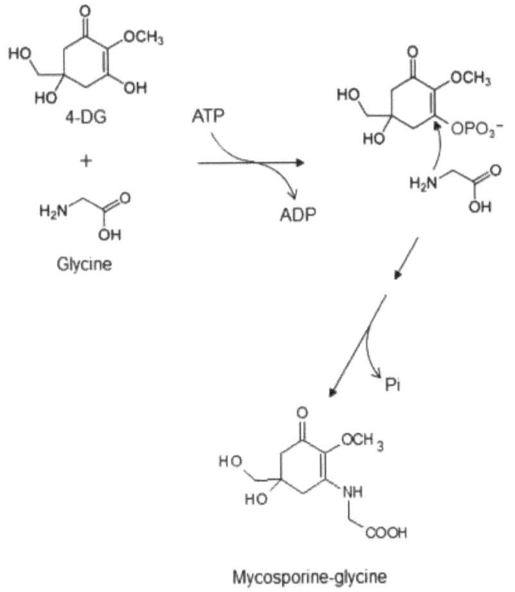

Fig. (6). Mycosporine-glycine biosynthesis reaction utilizing 4-DG as a substrate.

Subsequently, the replacement of the second amino acid at R_2 with mycosporine-glycine produces a bisubstituted MAA. In *A. variabilis* ATCC29413, the *Ava_3855* gene adjacent to the *Ava_3856* gene encodes the nonribosomal peptide synthetase (NRPS), which is the enzyme that catalyzes this substitution reaction*. The Ava_3855 protein catalyzes the condensation reaction of serine and mycosporine-glycine to produce shinorine. The Ava_3855 protein is composed of three domains: adenylation, thiolation, and thioesterization. (Fig. 7) outlines the reaction mechanism in which shinorine is produced from mycosporine-glycine. The carboxylate ion ($-COO^-$) of the serine is activated by adenylation by the adenylation domain of Ava_3855 and then covalently binds to the thiolation domain of the Ava_3855 protein. Next, the serine binds to the carbonyl group (R_1–C(=O)–R_2) at the C1 position of mycosporine-glycine. It is thought that the thioesterated domain is involved in the formation of imine (R_1–C(=NR_2)–R_3) at the C1 position in the final stage. As shown in (Fig. 7), the nitrogen atom in the serine residue is added to the C1 position by a 1,4-addition reaction to form shinorine.

* Depending on the cyanobacterial species, the protein corresponding to Ava_3855 is annotated as D-alanine-D-alanine ligase (D-Ala-D-Ala ligase) instead of NRPS.

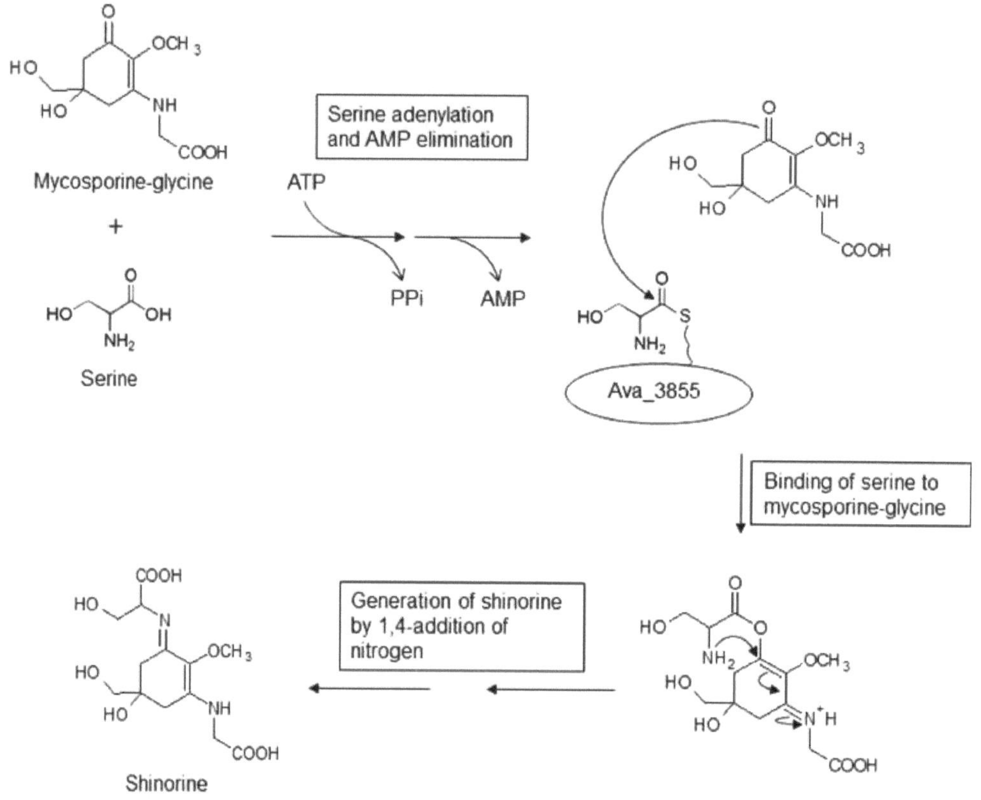

Fig. (7). Shinorine biosynthetic reaction involving mycosporine-glycine.

Gene Clusters for Biosynthesis of MAAs in Cyanobacteria

The shinorine biosynthetic genes *Ava_3858* (DHQS/EVS), *Ava_3857* (*O*–MT), *Ava_3856* (ATP-grasp), and *Ava_3855* (NRPS) in the *A. variabilis* ATCC29413 strain are adjacent to each other in the genome and form a gene cluster. As described above, 4-DG is first synthesized from intermediates of the primary metabolic pathways, the shikimate pathway, or the pentose phosphate pathway by the catalytic abilities of the Ava_3858 protein and the Ava_3857 protein. Subsequently, the Ava_3856 and Ava_3855 proteins biosynthesize shinorine by sequentially adding glycine and serine to 4-DG (Fig. **8**).

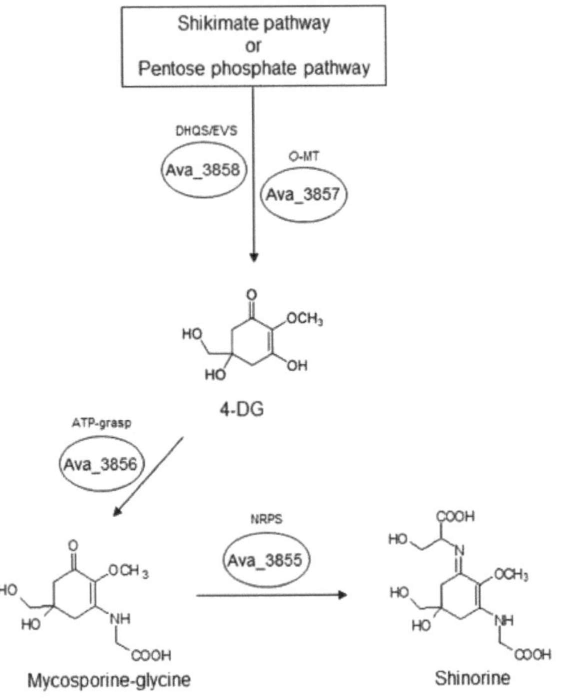

Fig. (8). Shinorine biosynthetic reaction involving mycosporine-glycine.

Because the MAA biosynthetic gene cluster in the *A. variabilis* ATCC29413 strain was reported first [4, 5], the gene cluster structure of this cyanobacterial strain tends to be recognized as the basic type. However, there are many strains with structures other than this cluster type. For example, in *Ava_3858-3855*, the genes are oriented in the same direction, but in the gene cluster consisting of *NpR5600*, *NpR5599*, *NpR5598*, and *NpF5597* in *N. punctiforme* ATCC29133, the direction of the *NpF5597* gene encoding D-Ala-D-Ala ligase is reversed (Fig. **9**) [5].

In *Halothece* sp. PCC7418, which biosynthesizes mycosporine-2-glycine, the three genes *PCC7418_1078* (*O*–MT), *PCC7418_1077* (ATP-grasp), and *PCC7418_1076* (D-Ala-D-Ala ligase) form a gene cluster. However, *PCC7418_1590*, which encodes DHQS, is located at a considerable distance on the genome and has an extra amino acid sequence in the N-terminal region compared to the DHQSs of other cyanobacteria (Fig. **9**) [8]. The function of this additional sequence of DHQS has not been clarified.

In *Cylindrospermum stagnale* PCC7417, a gene cluster consisting of 5 genes is formed. *ANS54016*, *ANS54017*, *ANS54018*, *ANS54019*, and *ANS54020* encode DHQS, O–MT, D-Ala-D-Ala ligase, ATP-grasp, and ATP-grasp, respectively.

The genes for ATP-grasp are duplicated, and the order of ATP-grasp and D-Al--D-Ala ligase is different from that of *A. variabilis* ATCC29413, *N. punctiforme* ATCC29133, and *Halothece* sp. PCC7418 (Fig. **9**). It was reported that when this gene cluster derived from *C. stagnale* PCC7417 was introduced into *E. coli*, the monosubstituted MAAs, mycosporine-lysine and mycosporine-ornithine, were biosynthesized [9]. Given these findings, it is considered that *C. stagnale* PCC7417 does not have the general type of MAA biosynthetic pathway shown in (Fig. **8**).

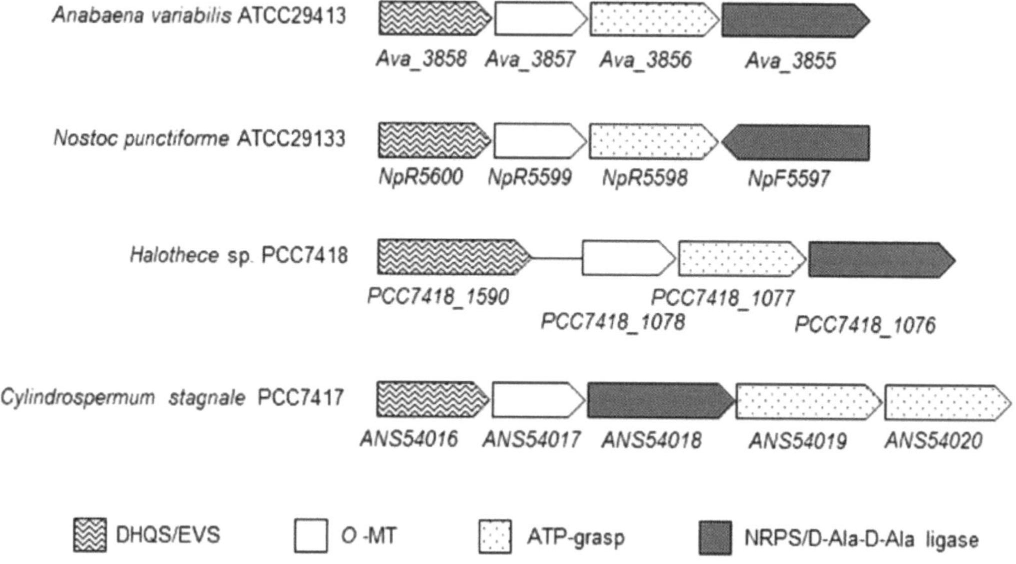

Fig. (9). MAA biosynthetic gene clusters of representitive cyanobacteria.

The MAA biosynthetic gene cluster containing the two ATP-grasp (MysC) genes found in *C. stagnale* PCC7417 has recently been shown to be specifically conserved in drought-tolerant cyanobacteria [10]. These ATP-grasps are phylogenetically classified into different groups (MysC2 and MysC3) apart from the typical one (MysC1). As shown in (Fig. **10**), MysC3 binds ornithine or lysine to 4-DG to produce mycosporine-ornithine or mycosporine-lysine. When mycosporine-ornithine is produced, isomers (isomer 1 and isomer 2) are generated depending on whether the site of ornithine bound to 4-DG is an α-amino group or a δ-amino group. MysC2 produces mycosporine-4-deoxygadusol-ornithine by binding the α-amino group of mycosporine-ornithine (isomer 2) to 4-DG. Finally, D-Ala-D-Ala ligase (MysD) combines mycosporine-ornithine (isomer 1) and mycosporine-4-deoxygadusol-ornithine to form mycosporine-2-(4-deoxygadu-ol-ornithine).

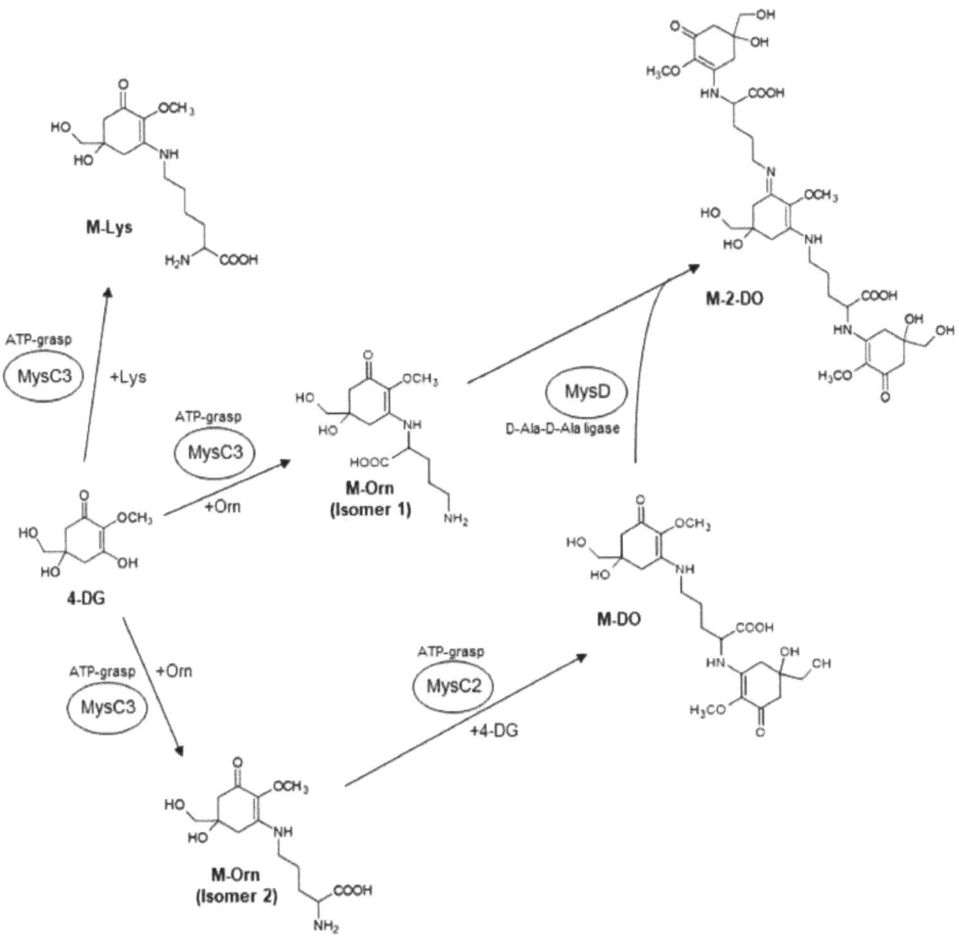

Fig. (10). MAA biosynthetic pathway in drought-tolerant cyanobacteria. M-Orn: mycosporine-ornithine, M-Lys: mycosporien-lysine, M-DO: mycosporine-4-deoxygadusol-ornithine, M-2-DO: mycosporine-2--4-deoxygadusol-ornithine).

The biosynthetic gene cluster for mycosporien-alanine biosynthesis found in *Hassallia byssoidea* collected from a stone monument in India also had a characteristic structure. Genes encoding DHQS (MysA), *O*–MT (MysB), and ATP-grasp (MysC) are present in the genome of this cyanobacterium. However, there was no gene encoding NRPS/D-Ala-D-Ala ligase. Instead, there was a gene encoding D-alanyl-D-alanine carboxypeptidase (MysD)*, and it is thought that this protein may be responsible for the conversion of mycosporine-glycine to mycosporine-alanine. A reaction to replace the glycine residue of mycosporine-glycine with alanine or a reaction to methylate the glycine residue has been proposed as the mechanism of mycosporine-alanine synthesis (Fig. **11**) [11].

* Note that in other cyanobacterial species, D-Ala-D-Ala ligase and NRPS are named MysD and MysE, respectively.

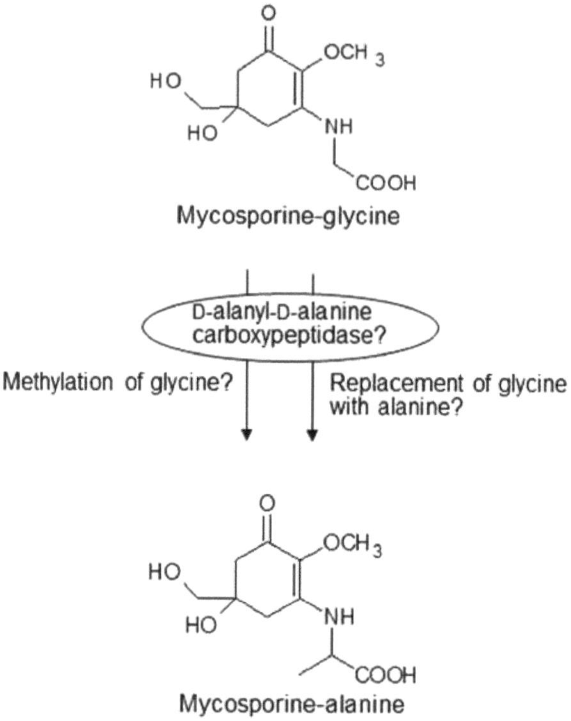

Mycosporine-glycine

D-alanyl-D-alanine carboxypeptidase?

Methylation of glycine?

Replacement of glycine with alanine?

Mycosporine-alanine

Fig. (11). Biosynthetic pathway of mycosporine-alanine in *Hassallia byssoidea*.

MAA Biosynthetic Genes in Species other than Cyanobacteria

In 2017, it was reported that red algae have a gene cluster with high homology to the MAA biosynthesis genes of cyanobacteria [12]. Six species of red algae (*Porphyra umbilicalis*, *Chondrus crispus*, *Cyanidioschyzon merolae*, *Galdieria sulphuraria*, *Porphyridium purpureum*, and *Pyropia yezoensis*) were investigated. As a result, it was clarified that three kinds of red algae (*P. umbilicalis*, *C. crispus*, and *P. yezoensis*) had a region encoding DHQS/EVS, *O*-MT, ATP-grasp, and D-Ala-D-Ala ligase. However, unlike cyanobacteria, DHQS/EVS and O-MT, and ATP-grasp and D-Ala-D-Ala ligase existed as fused genes (Fig. **12**). In addition, in the genomes of the closely related species *P. umbilicalis* and *P. yezo- ensis*, the two fusion genes were arranged with transcription directions pointing outward from one another, whereas in *C. crispus*, these genes were arranged in the opposite orientation. Cases in which certain genes were oriented differently in the MAA synthetic gene clusters were also confirmed in some cyanobacterial strains, including *N. punctiforme* ATCC29133 (Figs. **3–9**).

Dinoflagellates are an example of a fusion of MAA biosynthetic genes similar to red algae. In dinoflagellates capable of biosynthesizing MAAs, DHQS/EVS and *O*-MT exist as fused genes, whereas ATP-grasp and D-Ala-D-Ala ligase exist independently (Fig. **12**) [13]. It is possible that dinoflagellates acquired the fusion gene from red algae through secondary endosymbiosis [14].

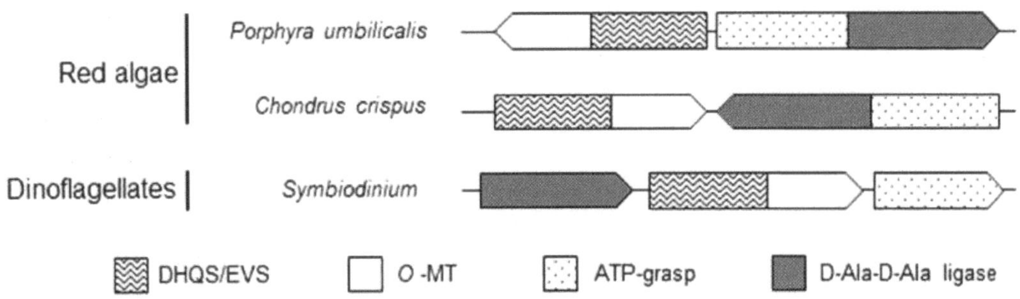

Fig. (12). MAA biosynthetic gene clusters of red algae and dinoflagellates.

Other organisms carrying the MAA biosynthetic genes include fungi, corals, and sea anemones. In fungi, the presence of DHQS, *O*-MT, and ATP-grasp has been confirmed, but NRPS/D-Ala-D-Ala ligase has not [5]. This is consistent with the fact that disubstituted MAAs have not been detected in fungi. However, in the coral *Acropora digitifera* and the sea anemone *Nematostella vectensis*, there is a full set of MAA biosynthetic genes (and DHQS and *O*–MT are fused together as in dinoflagellates) [9]. Thus, corals and sea anemones are thought to have the potential to biosynthesize MAAs. In the Gram-positive bacteruim *Actinomycetales*, genes encoding DHQS, *O*–MT, ATP-grasp, and D-Ala-D-Ala ligase are adjacently arranged on the genome, similar to cyanobacteria [15]. Although these bacteria carry complete MAA biosynthetic gene clusters, MAA accumulation has been undetectable (or detected only in trace amounts). By introducing the MAA biosynthetic gene cluster of *Actinomycetales* into the Gram-positive bacterium *Streptomyces*, biosynthesis of shinorine, porphyra-334, and mycosporine-glycine-alanine was detected* [15]. Given these findings, the MAA biosynthetic gene cluster of *Actinomycetales* has the function of MAA synthesis, but it is likely to be in a cryptonic state in this bacterium.

* Among these MAA molecules, mycosporine-glycine-alanine has not been detected in cyanobacteria.

MAA Biosynthetic Genes as Gene Resources

The MAA biosynthetic genes found mainly in cyanobacteria can be used as gene resources for producing MAAs in other species. Table **1** shows the MAA biosynthetic genes of cyanobacterial strains, including *Anabaena variabilis*

ATCC29413 [5], *Halothece* sp. PCC7418 [8], *Scytonema cf. crispum* [15], *C. stagnale* PCC7417 [9], *Nostoc linckia* NIES-25 [16], and *N. flagelliforme* [10], that were introduced into *E. coli* and the MAAs detected in the cells.

Table 1. Introduction of cyanobacterial MAA biosynthetic genes into *E. coli*

Cyanobacterial Strain (Biosynthesized MAAs)	MAA Biosynthetic Genes Introduced	MAAs Accumulated in Transformed *Escherichia coli* Cells
Anabaena variabilis ATCC29413 (Shinorine) [5]	*Ava_3858* (DHQS/EVS) *Ava_3857* (*O*–MT) *Ava_3856* (ATP-grasp) *Ava_3855* (D-Ala-D-Ala ligase)	Shinorine
Halothece sp. PCC7418 (Mycorsporine-2-glycine) [8]	*PCC7418_1590* (DHQS/EVS) *PCC7418_1078* (*O*–MT) *PCC7418_1077* (ATP-grasp) *PCC7418_1076* (D-Al--D-Ala ligase)	Mycosporine-2-glycine
Scytonema cf. Crispum (Shinorine, Palythine-serine, Pentose-bound shinorine, Pentose-bound palythine-serine) [17]	*MysA* (DHQS) *MysB* (*O*–MT) *MysC* (ATP-grasp) *MysE* (NRPS)	Shinorine
Cylindrospermum stagnale PCC7417 (Not reported) [9]	*ANS54016* (DHQS) *ANS54017* (*O*–MT) *ANS54018* (D-Ala-D-Ala ligase) *ANS54019* (ATP-grasp) *ANS54020* (ATP-grasp)	Mycosporine-lysine Mycosporine-ornithine
Nostoc linckia NIES-25 (Shinorine) [16]	*NIES25_64110* (Phytanoyl-CoA dioxygenase) *NIES25_64130* (EVS) *NIES25_64140* (*O*–MT) *NIES25_64150* (ATP-grasp) *NIES25_64160* (D-Ala-D-Ala ligase)	4-DG Mycosporine-glycine Shinorine Porphyra-334 Mycosporine-glycine-alanine Palythine-serine Palythine-threonine Palythine-alanine
Nostoc flagelliforme (Mycosporine-2-(4-deoxygadusol-ornithine)) [10, 18]	*MysA* (DHQS/EVS) *MysB* (*O*–MT) *MysD* (D-Ala-D-Ala ligase) *MysC2* (ATP-grasp) *MysC3*(ATP-grasp)	4-DG Mycosporine-lysine Mycosporine-ornithine Mycosporine-4-deoxygadusol-ornithine (from the introduction of *MysABC2C3*)

When the MAA biosynthetic gene cluster of *A. variabilis* ATCC29413 was introduced into *E. coli*, shinorine was biosynthesized. When the gene cluster of *Halothece* sp. PCC7418 was introduced, mycosporine-2-glycine was produced in *E. coli* cells. Thus, the MAAs biosynthesized in the host *E. coli* cells were the same as those biosynthesized in the cyanobacterium with the original gene cluster. In contrast, some reports showed that the types of MAAs biosynthesized differed between cyanobacteria and host *E. coli*. When the gene cluster of *Scytonema cf. crispum* was introduced, shinorine was biosynthesized in the host *E. coli* cells. Interestingly, however, palythine-serine, hexose-bound shinorine, and hexose-bound palythine-serine were detected together with shinorine in *Scytonema cf. crispum* itself [17]. This suggests that genes involved in the biosynthesis of palythine-serine and hexose-bound MAAs existed in the genome of *Scytonema cf. crispum* outside of the introduced gene cluster. In *N. linckia* NIES-25, phytanoyl-CoA dioxygenase, which is located near the EVS gene, was discovered as a novel MAA biosynthetic gene [16]. Phytanoyl-CoA dioxygenase is thought to decarboxylate and demethylate the glycine residues of shinorine, porphyra-334, and mycosporine-glycine-alanine and convert them to palythine-serine, palythine-threonine, and palythine-alanine, respectively (Fig. **13**).

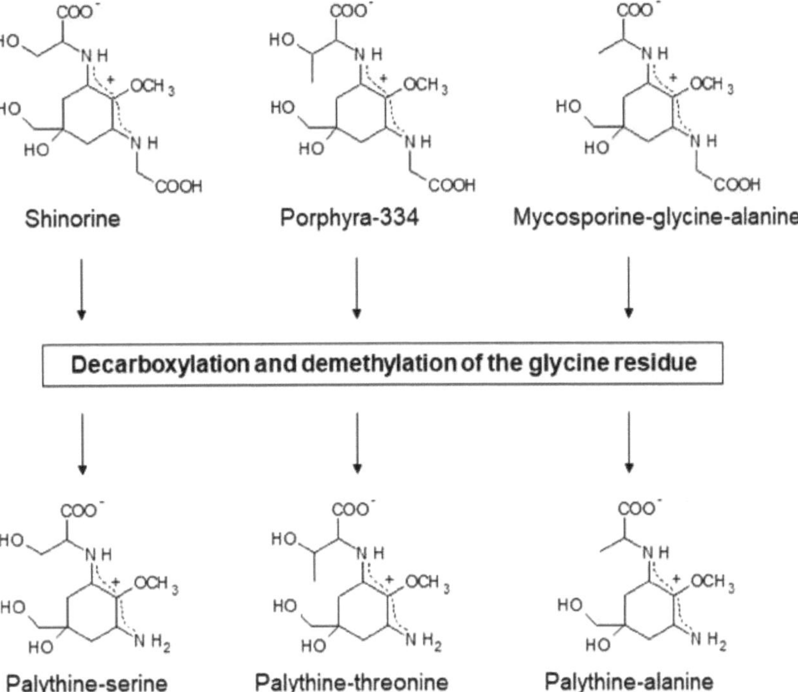

Fig. (13). Conversion of shinorine, porphyra-334, and mycosporine-glycine-alanine to palythine-serine, palythine-threonine, and palythine-alanine, respectively, by phytanoyl-CoA dioxygenase.

Substrate Specificity of NRPS/D-Ala-D-Ala Ligase

In the common MAA biosynthetic pathway in cyanobacteria, disubstituted MAAs are synthesized from mycosporine-glycine (Fig. **8**). The most frequently detected MAA in cyanobacteria is shinorine, but various types of disubstituted MAAs can accumulate in cells depending on the cyanobacterial strain. It is thought that NRPS and D-Ala-D-Ala ligase contribute to the diversity of MAA biosynthesis. For example, when a recombinant protein of NIES25_64160 (MysD), which is the D-Ala-D-Ala ligase of *N. linckia* NIES-25, was incubated with 20 different amino acids in the presence of mycospoeine-glycine and ATP *in vitro*, the enzyme reacted with only 6 kinds of amino acids, alanine, arginine, cysteine, glycine, serine, and threonine, and generated mycosporine-glycine-alanine, mycosporine-glycine-arginine, mycosporine-glycine-cysteine, mycosporine-2-glycine, and shinorine, porphyra-334, respectively* [16]. In addition, it was reported that the Ava_3855 protein (NRPS), which synthesizes shinorine by adding serine to mycosporine-glycine in *A. variabilis* ATCC29413, could not synthesize porphyra-334 when reacted with threonine instead of serine in *in vitro* experiments [5]. However, if the mechanism of substrate specificity of NRPS/D-Ala-D-Ala ligase was known at the protein structure level, a preferred disubstituted MAAs could be generated by utilizing NRPS/D-Ala-D-Ala ligase with a designed amino acid sequence. MAAa have different activities depending on their molecular structures. Because it is useful to prepare MAA molecules with a structure most suitable for the application, elucidation of the substrate specificity of NRPS/D-Ala-D-Ala ligase would be a very interesting research subject.

* Among these six amino acids, serine and threonine were significantly more reactive than other amino acids and had a high affinity for NIES25_64160 (MysD).

REGURATORY MECHANISMS OF MAA BIOSYNTHETIC PATHWAYS

It is known that the accumulation levels of MAAs in living organisms fluctuate under the influence of various environmental factors. However, the molecular mechanisms such as gene expression and post-translational regulation of enzymes involved in MAA biosynthetic pathways have not been clarified. Here we detail the effects of abiotic stresses on the accumulation of MAAs, focusing on cyanobacteria.

UV Irradiation Stress

UV irradiation is generally known to promote the accumulation of MAAs. This indirectly indicates that MAAs play a part in UV protection *in vivo*. From research reports on cyanobacteria, it was shown that UV-B irradiation had a greater effect than UV-A irradiation. As described in Chapter 2, UV-B irradiation caused the induction of MAAs in many cyanobacterial strains, such as *Anabaena doliolum* [19], *Anabaena variabilis* ATCC29413 [20], *Anabaena* sp [21], *Nodularia* [22], *Chlorogloeopsis* PCC6912 [23], *Nostoc commune* [21], *Scytonema* sp [21], *Halothece* sp. PCC7418 [8], and *Nostoc flagelliforme* [18]. In *N. flagelliforme*, the response of the MAA biosynthetic gene cluster (*MysA-Mys--MysD-MysC2-MysC3*) to UV-B irradiation was investigated by reverse transcription polymerase chain reaction (RT-PCR) analysis. As a result, it was clarified that the expression of all five genes in the cluster was upregulated by UV-B irradiation [18]. The UV-B responsive promoter region located upstream of this MAA biosynthetic gene cluster has also been roughly determined [18]. In addition to cyanobacteria, induction of MAA accumulation by UV irradiation has been confirmed in yeast, macroalgae, haptophyta, diatoms, and marine microalgae such as dinoflagellates [24 - 27]. Pterin (Fig. **14**) has been proposed as a candidate factor of UV-B receptors in cyanobacteria [28]. However, so far no association has been found between pterin and UV-B irradiation signal transduction pathways or MAA biosynthetic gene expression [29]. In *N. commune*, UV irradiation has been shown to promote biosynthesis and secretion of water stress proteins [30]. Water stress proteins are thought to play an important role in the transport and diffusion of UV-absorbing substances, including MAAs and scytonemin, by changing the three-dimensional conformation of the extracellular matrix in *N. commune*.

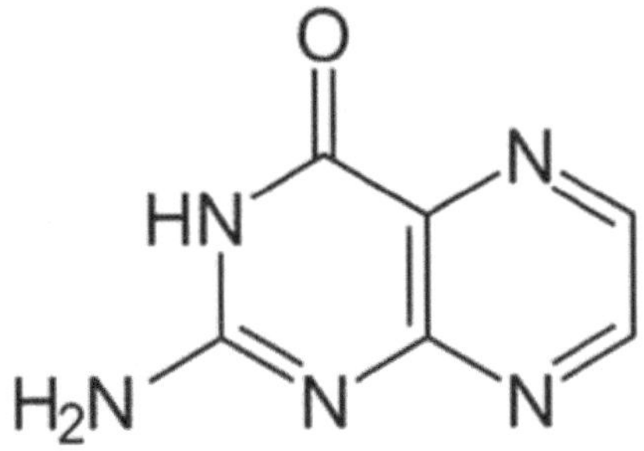

Fig. (14). Molecular structure of pterin.

Salt Stress and Osmotic Stress

In the halotolerant cyanobacterium *Halothece* sp. PCC7418, NaCl stress has been shown to increase the accumulation of mycosporine-2-glycine [8]. In addition, the expression profiles of the biosynthetic genes for mycosporine-2-glycine production under NaCl stress have been investigated. RT-PCR analysis revealed that the expression levels of all four mycosporine-2-glycine biosynthetic genes, *PCC7418_1590* (DHQS), *PCC7418_1078* (*O*–MT), *PCC7418_1077* (ATP-grasp), and *PCC7418_1076* (D-Ala-D-Ala ligase), were significantly increased by

NaCl stress. As mentioned above, not all of these genes are adjacent on the genome, but their positions are separated by *PCC7418_1590* and *PCC7418_1078-1076* (Fig. **9**). By introducing DNA fragments that contain the ORFs of all four genes along with hundreds of bases upstream of the *PCC7418_1590* gene and the *PCC7418_1078* gene, mycosporine-2-glycine can be biosynthesized in *E. coli* (Fig. **15**). It was confirmed that the biosynthesis of mycosporine-2-glycine was induced by applying NaCl stress to this transformed *E. coli* cells. In addition, the accumulations of DHQS and D-Ala-D-Ala ligase increased [8]. Therefore, it is considered that the upstream regions of the *PCC7418_1590* gene and the *PCC7418_1078* gene contain NaCl stress-responsive promoters and transcriptional regulatory sequences. In addition to cyanobacteria, accumulation of MAAs has been shown to increase with increasing seawater salinity in the marine dinophyceae *Gymnodinium catenatum* [31]. Furthermore, the red alga *Porphyra columbina* is known to increase the accumulation of MAAs in cultures with added ammonium salt [32].

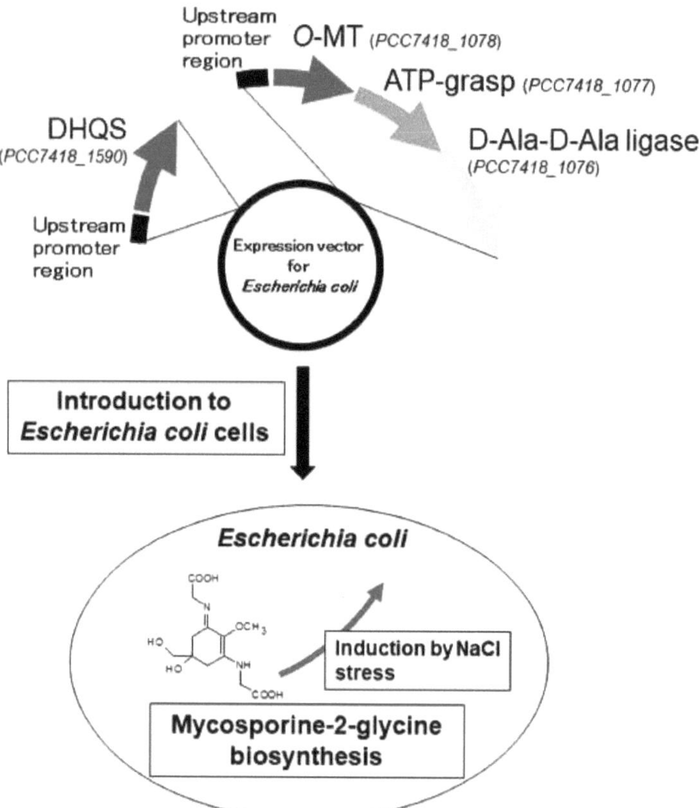

Fig. (15). Production of mycosporine-2-glycine in *Escherichia coli* after the introduction of MAA biosynthetic genes derived from *Halothece* sp. PCC7418 strain.

Other Abiotic Factors

This section describes the effects of abiotic factors other than UV irradiation, salt stress, or osmotic stress on the biosynthesis of MAAs.

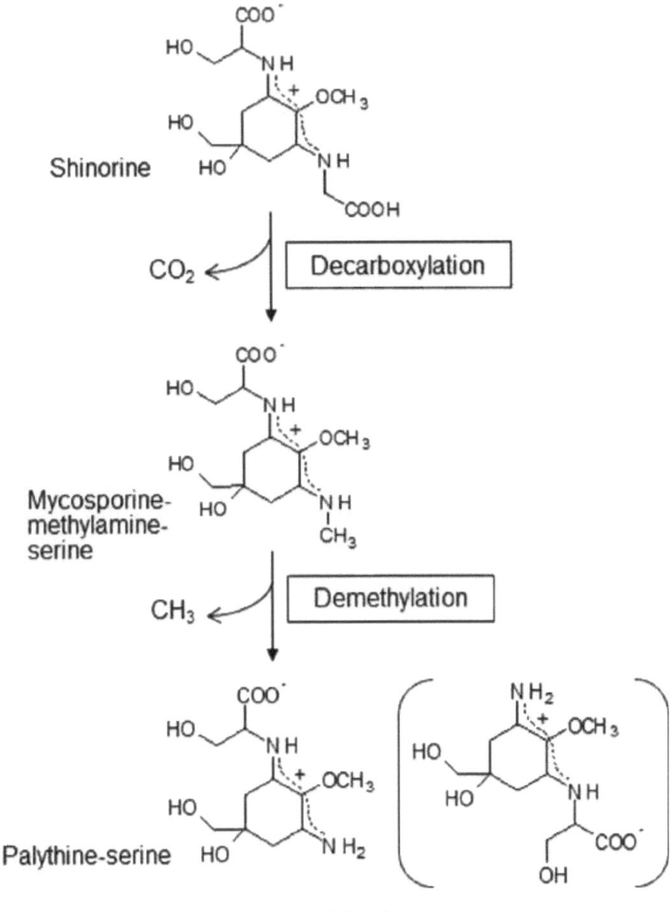

Fig. (16). Palythine-serine biosynthetic pathway from shinorine.

Nutrient Concentration

The nutrient concentration in the surrounding environment also affects the biosynthesis of MAAs. For example, in the halotolerant cyanobacterium *Halothece* sp. PCC7418, it was reported that oversupply of nitrate affected the amount of mycosporine-2-glycine biosynthesis [33]. In addition, sulfur deficiency altered the composition of MAAs in the cyanobacterium *Anabaena variabilis* ATCC29413 strain [34]. This cyanobacterium normally accumulated only shinorine, but when sulfur was deficient, it began to produce palythine-serine. Palythine-serine is thought to be produced by decarboxylation and demethylation of shinorine (Fig. **16**) [35]. Glycine decarboxylase has been proposed as an enzyme responsible for the conversion of shinorine to palythine-serine in cyanobacterial cells*. The addition of methionine, a sulfur-containing amino acid, to sulfur-deficient culture medium stopped the biosynthesis of palythine-serine [34]. Given this result, the shinorine-derived methyl group ($-CH_3$) may be involved in methionine regeneration under sulfur-deficient conditions. Methionine is produced by the transfer of a methyl group derived from 5-methyltetra hydrofolate (5-methyl THF) to homocysteine (Fig. **17**). It is known that under sulfur-deficient conditions, the intracellular concentration of methionine and its derivative S-adenosylmethionine (SAM) decreases, while the concentration of homocysteine increases. Given these observations, it is presumed that the following reaction mechanism occurs under sulfur-deficient conditions. When sulfur is deficient and the concentration of methionine decreases, the methyl group derived from shinorine methylates tetrahydrofolate (THF) to produce 5-methyl THF (Fig. **17**). Then, the methyl group of 5-methyl THF is transfered to homocysteine and used for the production of methionine.

* As mentioned above, it was experimentally shown that phytanoyl-CoA dioxygenase is the enzyme responsible for this reaction. The results show that by introducing phytanoyl-CoA dioxygenase into *E. coli* cells, shinorine, porphyra-334, and mycosporine-glycine-alanine are converted to palythine-serine, palythine-threonine, and palythine-alanine, respectively.

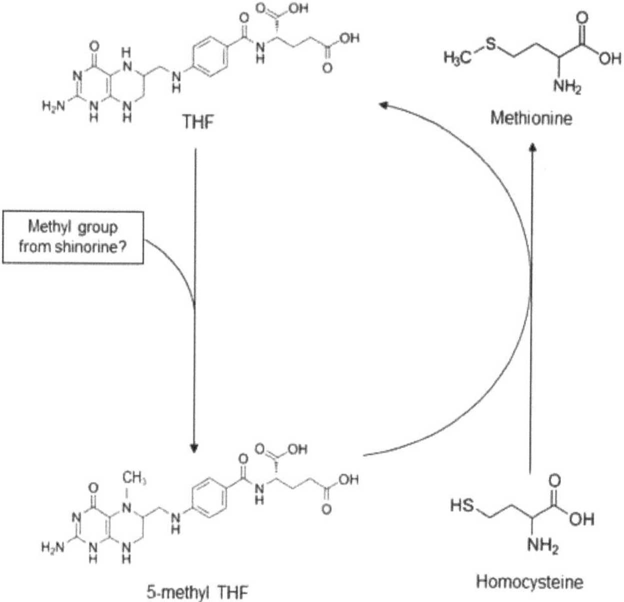

Fig. (17). Regeneration of methionine from homocysteine.

Temperature

As mentioned above, there are reports that temperature changes affected the amount of MAAs accumulated in cyanobacteria and corals. However, the regulatory mechanism is unknown, and more detailed analysis is required in the future.

Far-red Light

It has been reported that far-red light (FR) may affect the biosynthesis of MAAs in cyanobacteria [36]. Far-red light generally represents light in the wavelength range of 700 to 800 nm. According to that report, the expression of MAA biosynthetic genes of the filamentous cyanobacterium *Chlorogloeopsis fritschii* PCC6912 was increased not only by UV irradiation but also by FR irradiation. Increased accumulation of shinorine and mycosporine-glycine was also confirmed under both conditions.

Bottleneck Factors in MAA Synthesis

In MAA-producing organisms, there are several factors that are thought to be bottlenecks in biosynthesis of MAAs.

Supply of Precursors of MAAs

The most fundamental is the supply of precursors *via* the shikimate or pentose phosphate pathways. Various basic metabolic mechanisms are involved in the synthesis of precursor molecules of MAAs, and it is desired to clarify these networks.

Unraveling the molecular control of MAA biosynthesis would lead to industrial production of MAAs. For example, introduction of MAA biosynthetic genes into a suitable microorganism could enable efficient production of MAAs. Fom the viewpoint of industrial production of MAAs, it is important to create an *in vivo* pathway that can sufficiently supply the substance that becomes a bottleneck in the synthesis of MAA by genetic manipulation.

Supply of Amino Acids

The supply of amino acids used in the final MAA synthesis process would be also essential. The synthesis of MAAs might compete with the synthesis of proteins and other substances for amino acids. For example, the halotolerant cyanobacterium *Halothece* sp. PCC7418 promotes the biosynthesis of mycosporine-2-glycine under salt stress conditions, while simultaneously producing large amounts of glycine betaine as a compatible solute [8, 33]. Since glycine betaine is synthesized by methylation of glycine, glycine betaine biosynthesis and mycosporine-2-glycine biosynthesis compete for glycine. In fact, in *Halothece* sp. PCC7418, the amount of mycosporine-2-glycine synthesized is only about 1/1000 of that of glycine betaine [33].

LOCALIZATION OF MAAS

MAAs are basically localized in the cytoplasm, but there have been reports in which they existed extracellularly. In the unicellular freshwater cyanobacterium *Microcystis aeruginosa* PCC7806, shinorine was localized in the extracellular polysaccharide (EPS) matrix. It was suggested that shinorine was involved in EPS formation and cell–cell interaction in this cyanobacterium [37]. In diatoms, it has been reported that shinorine, porphyra-334, palythine, asterina-330, palythinol, and palythinic acid were present in the frustule structure covering the cells [38]. Besides, it has been reported that MAAs were secreted or eluted into the surrounding water during algal blooms in the cyanobacteriaum *Trichodesmium* spp. [39] and in the dinoflagellates *Lingulodinium polyedra* [40] and *Prorocentrum micans* [41].

CONCLUDING REMARKS

In this chapter, the reaction mechanisms for the synthetic pathways of MAAs were presented based mainly on results from cyanobacteria. Although these biosynthetic mechanisms have been roughly clarified, there remain many points to be clarified regarding the detailed regulatory mechanisms. In particular, to achieve industrial production of MAAs, it would be important to elucidate the regulatory mechanisms of the expression of MAA biosynthetic genes and to clarify the substrate specificity of NRPS/D-Ala-D-Ala ligase.

LIST OF ABBREVIATIONS

4-DG	4-deoxygadusol
ATP	Adenosine triphosphate
DAHP	3-deoxy-D-arabinohepturonic acid 7-phosphate
DDG	Demethyl-4-deoxygadusol
DHQ	3-dehydroquinate
DHQS	3-dehydroquinate synthase
EPS	Extracellular polysaccharide matrix
EVS	2-*epi*-5-*epi*-valiolone synthase
GABA	Gamma-aminobutyric acid
G3P	Glyceraldehyde-3-phosphate
G6P	Glucose-6-phosphate

MAA	Mycosporine-like amino acid
NADPH/NADP	Nicotinamide adenine dinucleotide phosphate
NRPS	Nonribosomal peptide synthetase (NRPS)
O MT	*O*–methyltransferase
ROS	Reactive oxygen species
UV	Ultraviolet

CONSENT FOR PUBLICATION

Not applicable.

CONFLICT OF INTEREST

The author declares no conflict of interest, financial or otherwise.

ACKNOWLEDGEMENTS

Declared none.

REFERENCES

[1] Favre-Bonvin, J.; Bernillon, J.; Salin, N.; Arpin, N. Biosynthesis of mycosporines: Mycosporine glutaminol in *Trichothecium roseum. Phytochemistry,* **1987**, *26*(9), 2509-2514.

[2] Portwichi, A.; Garcia-Pichel, F. Biosynthetic pathway of mycosporines (mycosporine-like amino acids) in the cyanobacterium *Chlorogloeopsis* sp. strain PCC 6912. *Phycologia,* **2003**, *42*(4), 384-392.

[3] Shick, J.M.; Romaine-Lioud, S. Ferrier-Page`s, C.; Gattuso, J.-P. Ultraviolet-B radiation stimulates shikimate pathway-dependent accumulation of mycosporine-like amino acids in the coral *Stylophora pistillata* despite decreases in its population of symbiotic dinoflagellates. *Limnol. Oceanogr.,* **1999**, *44*(7), 1667-1682.

[4] Singh, S.P.; Klisch, M.; Sinha, R.P.; Häder, D.P. Genome mining of mycosporine-like amino acid (MAA) synthesizing and non-synthesizing cyanobacteria: A bioinformatics study. *Genomics,* **2010**, *95*(2), 120-128. [From NLM.]. [http://dx.doi.org/10.1016/j.ygeno.2009.10.002]

[5] Balskus, E.P.; Walsh, C.T. The genetic and molecular basis for sunscreen biosynthesis in cyanobacteria. *Science,* **2010**, *329,* 1653-1656.

[6] Spence, E.; Dunlap, W.C.; Shick, J.M.; Long, P.F. Redundant pathways of sunscreen biosynthesis in a cyanobacterium. *ChemBioChem,* **2012**, *13*(4), 531-533. [http://dx.doi.org/10.1002/cbic.201100737]

[7] Pope, M.A.; Spence, E.; Seralvo, V.; Gacesa, R.; Heidelberger, S.; Weston, A.J.; Dunlap, W.C.; Shick, J.M.; Long, P.F. *O*-Methyltransferase is shared between the pentose phosphate and shikimate pathways and is essential for mycosporine-like amino acid biosynthesis in *Anabaena variabilis* ATCC 29413. *ChemBioChem,* **2015**, *16*(2), 320-327. [http://dx.doi.org/10.1002/cbic.201402516]

[8] Waditee-Sirisattha, R.; Kageyama, H.; Sopun, W.; Tanaka, Y.; Takabe, T. Identification and upregulation of biosynthetic genes required for accumulation of Mycosporine-2-glycine under salt stress conditions in the halotolerant cyanobacterium *Aphanothece halophytica. Appl. Environ.*

Microbiol., **2014**, *80*(5), 1763-1769.
[http://dx.doi.org/10.1128/AEM.03729-13]

[9] Katoch, M.; Mazmouz, R.; Chau, R.; Pearson, L.A.; Pickford, R.; Neilan, B.A. Heterologous production of cyanobacterial mycosporine-like amino acids mycosporine-ornithine and mycosporine-lysine in *Escherichia coli. Appl. Environ. Microbiol.,* **2016**, *82*(20), 6167-6173.
[http://dx.doi.org/10.1128/AEM.01632-16]

[10] Zhang, Z.C.; Wang, K.; Hao, F.H.; Shang, J.L.; Tang, H.R.; Qiu, B.S. New types of ATP-grasp ligase are associated with the novel pathway for complicated mycosporine-like amino acid production in desiccation-tolerant cyanobacteria. *Environ. Microbiol.,* **2021**, *23*(11), 6420-6432.
[http://dx.doi.org/10.1111/1462-2920.15732]

[11] Saha, S.; Sen, A.; Mandal, S.; Adhikary, S.P.; Rath, J. Mycosporine-alanine, an oxo-mycosporine, protect *Hassallia byssoidea* from high UV and solar irradiation on the stone monument of Konark. *J. Photochem. Photobiol. B,* **2021**, *224*, 112302.
[http://dx.doi.org/10.1016/j.jphotobiol.2021.112302]

[12] Brawley, S.H.; Blouin, N.A.; Ficko-Blean, E.; Wheeler, G.L.; Lohr, M.; Goodson, H.V.; Jenkins, J.W.; Blaby-Haas, C.E.; Helliwell, K.E.; Chan, C.X. Insights into the red algae and eukaryotic evolution from the genome of *Porphyra umbilicalis* (Bangiophyceae, Rhodophyta). *Proc. Natl. Acad. Sci. USA,* **2017**, *114*(31), E6361-E6370.
[http://dx.doi.org/10.1073/pnas.1703088114]

[13] Shoguchi, E.; Beedessee, G.; Tada, I.; Hisata, K.; Kawashima, T.; Takeuchi, T.; Arakaki, N.; Fujie, M.; Koyanagi, R.; Roy, M.C. Two divergent Symbiodinium genomes reveal conservation of a gene cluster for sunscreen biosynthesis and recently lost genes. *BMC Genomics,* **2018**, *19*(1), 458.
[http://dx.doi.org/10.1186/s12864-018-4857-9]

[14] Shoguchi, E. Gene clusters for biosynthesis of mycosporine-like amino acids in dinoflagellate nuclear genomes: Possible recent horizontal gene transfer between species of Symbiodiniaceae (Dinophyceae). *J. Phycol.,* **2021**.
[http://dx.doi.org/10.1111/jpy.13219]

[15] Miyamoto, K.T.; Komatsu, M.; Ikeda, H. Discovery of gene cluster for mycosporine-like amino acid biosynthesis from *Actinomycetales* microorganisms and production of a novel mycosporine-like amino acid by heterologous expression. *Appl. Environ. Microbiol.,* **2014**, *80*(16), 5028-5036.
[http://dx.doi.org/10.1128/AEM.00727-14]

[16] Chen, M.; Rubin, G.M.; Jiang, G.; Raad, Z.; Ding, Y. Biosynthesis and heterologous production of mycosporine-like amino acid palythines. *J. Org. Chem.,* **2021**, *86*(16), 11160-11168.
[http://dx.doi.org/10.1021/acs.joc.1c00368]

[17] D'Agostino, P.M.; Javalkote, V.S.; Mazmouz, R.; Pickford, R.; Puranik, P.R.; Neilan, B.A. Comparative profiling and discovery of novel glycosylated mycosporine-like amino acids in two strains of the cyanobacterium *Scytonema* cf. *crispum. Appl. Environ. Microbiol.,* **2016**, *82*(19), 5951-5959.
[http://dx.doi.org/10.1128/AEM.01633-16]

[18] Shang, J.L.; Zhang, Z.C.; Yin, X.Y.; Chen, M.; Hao, F.H.; Wang, K.; Feng, J.L.; Xu, H.F.; Yin, Y.C.; Tang, H.R. UV-B induced biosynthesis of a novel sunscreen compound in solar radiation and desiccation tolerant cyanobacteria. *Environ. Microbiol.,* **2018**, *20*(1), 200-213.
[http://dx.doi.org/10.1111/1462-2920.13972]

[19] Singh, S.P.; Sinha, R.P.; Klisch, M.; Hader, D.P. Mycosporine-like amino acids (MAAs) profile of a rice-field cyanobacterium *Anabaena doliolum* as influenced by PAR and UVR. *Planta,* **2008**, *229*(1), 225-233.
[http://dx.doi.org/10.1007/s00425-008-0822-1]

[20] Singh, S.P.; Klisch, M.; Sinha, R.P.; Ha¨der, D-P. Effects of abiotic stressors on synthesis of the mycosporine-like amino acid shinorine in the cyanobacterium *Anabaena variabilis* PCC 7937.

Photochem. Photobiol., **2008**, *84*, 1500-1505.

[21] Sinha, R.P.; Klisch, M.; Helbling, E.W.; Hader, D-P. Induction of mycosporine-like amino acids (MAAs) in cyanobacteria by solar ultraviolet-B radiation. *J. Photochem. Photobiol. B,* **2001**, *60*, 129-135.

[22] Sinha, R.P.; Ambasht, N.K.; Sinha, J.P.; Klisch, M.; Hader, D.P. UV-B-induced synthesis of mycosporine-like amino acids in three strains of *Nodularia* (cyanobacteria). *J. Photochem. Photobiol. B,* **2003**, *71*(1-3), 51-58.
[http://dx.doi.org/10.1016/j.jphotobiol.2003.07.003]

[23] Portwich, A.; Garcia-Pichel, F. Ultraviolet and osmotic stresses induce and regulate the synthesis of mycosporines in the cyanobacterium *Chlorogloeopsis* PCC 6912. *Arch. Microbiol.,* **1999**, *172*, 187-192.

[24] Libkind, D.; Pérez, P.; Sommaruga, R.; Diéguez Mdel, C.; Ferraro, M.; Brizzio, S.; Zagarese, H.; van Broock, M. Constitutive and UV-inducible synthesis of photoprotective compounds (carotenoids and mycosporines) by freshwater yeasts. *Photochem. Photobiol. Sci.,* **2004**, *3*(3), 281-286. [From NLM.].
[http://dx.doi.org/10.1039/b310608j]

[25] Riegger, L.; Robinson, D. Photoinduction of UV-absorbing compounds in *Antarctic diatoms* and Phaeocystis antarctica. *Mar. Ecol. Prog. Ser.,* **1997**, *160*, 13-25.

[26] Klisch, M.; Häder, D.P. Mycosporine-like amino acids in the marine dinoflagellate *Gyrodinium dorsum*: induction by ultraviolet irradiation. *J. Photochem. Photobiol. B,* **2000**, *55*(2-3), 178-182. [From NLM.].
[http://dx.doi.org/10.1016/s1011-1344(00)00045-2]

[27] Arróniz-Crespo, M.; Sinha, R.P.; Martínez-Abaigar, J.; Núñez-Olivera, E.; Häder, D.P. Ultraviolet radiation-induced changes in mycosporine-like amino acids and physiological variables in the red alga *Lemanea fluviatilis. J. Freshwat. Ecol.,* **2005**, *20*(4), 677-687.

[28] Portwich, A.; Garcia-Pichel, F. A novel prokaryotic UVB photoreceptor in the cyanobacterium *Chlorogloeopsis* PCC 6912. *Photochem. Photobiol.,* **2000**, *71*(4), 493-498. [From NLM.].
[http://dx.doi.org/10.1562/0031-8655(2000)071<0493:anpupi>2.0.co;2]

[29] Gao, Q.; Garcia-Pichel, F. Microbial ultraviolet sunscreens. *Nat. Rev. Microbiol.,* **2011**, *9*(11), 791-802.
[http://dx.doi.org/10.1038/nrmicro2649]

[30] Wright, D.J.; Smith, S.C.; Joardar, V.; Scherer, S.; Jervis, J.; Warren, A.; Helm, R.F.; Potts, M. UV irradiation and desiccation modulate the three-dimensional extracellular matrix of *Nostoc commune* (Cyanobacteria). *J. Biol. Chem.,* **2005**, *280*(48), 40271-40281.
[http://dx.doi.org/10.1074/jbc.M505961200]

[31] Vale, P. Effects of light and salinity stresses in production of mycosporine-like amino acids by *Gymnodinium catenatum* (Dinophyceae). *Photochem. Photobiol.,* **2015**, *91*(5), 1112-1122.
[http://dx.doi.org/10.1111/php.12488]

[32] Peinado, N.K.; Díaz, R.T.A.; Figueroa, F.L.; Helbling, E.W. Ammonium and UV radiation stimulate the accumulation of mycosporine-like amino acids in *Porphyra columbina* (Rhodophyta) from Patagonia, Argentina. *J. Phycol.,* **2004**, *40*(2), 248-259.

[33] Waditee-Sirisattha, R.; Kageyama, H.; Fukaya, M.; Rai, V.; Takabe, T. Nitrate and amino acid availability affects glycine betaine and mycosporine-2-glycine in response to changes of salinity in a halotolerant cyanobacterium *Aphanothece halophytica. FEMS Microbiol. Lett.,* **2015**, *362*(23), fnv198.
[http://dx.doi.org/10.1093/femsle/fnv198]

[34] Singh, S.P.; Klisch, M.; Sinha, R.P.; Hader, D-P. Sulfur deficiency changes mycosporine-like amino acid (MAA) composition of *Anabaena variabilis* PCC 7937: A possible role of sulfur in MAA bioconversion. *Photochem. Photobiol.,* **2010**, *86*, 862-870.

[35] Carreto, J.I.; Carignan, M.O.; Montoya, N.G. A high-resolution reverse-phase liquid chromatography

method for the analysis of mycosporine-like amino acids (MAAs) in marine organisms. *Mar. Biol.,* **2005**, *146*, 237-252.

[36] Llewellyn, C.A.; Greig, C.; Silkina, A.; Kultschar, B.; Hitchings, M.D.; Farnham, G. Mycosporine-like amino acid and aromatic amino acid transcriptome response to UV and far-red light in the cyanobacterium *Chlorogloeopsis fritschii* PCC 6912. *Sci. Rep.,* **2020**, *10*(1), 20638. [http://dx.doi.org/10.1038/s41598-020-77402-6]

[37] Hu, C.; Völler, G.; Süßmuth, R.; Dittmann, E.; Kehr, J.C. Functional assessment of mycosporine-like amino acids in *Microcystis aeruginosa* strain PCC 7806. *Environ. Microbiol.,* **2015**, *17*(5), 1548-1559. [From NLM.]. [http://dx.doi.org/10.1111/1462-2920.12577]

[38] Ingalls, A.E.; Whitehead, K.; Bridoux, M.C. Tinted windows: The presence of the UV absorbing compounds called mycosporine-like amino acids embedded in the frustules of marine diatoms. *Geochim. Cosmochim. Acta,* **2010**, *74*(1), 104-115. [http://dx.doi.org/10.1016/j.gca.2009.09.012]

[39] Subramaniam, A.; Carpenter, E.J.; Karentz, D.; Falkowski, P.G. Bio-optical properties of the marine diazotrophic cyanobacteria *Trichodesmium* spp. I. Absorption and photosynthetic action spectra. *Limnol. Oceanogr.,* **1999**, *44*(3), 608-617.

[40] Vernet, M.; Whitehead, K. Release of ultraviolet-absorbing compounds by the red-tide dinoflagellate *Lingulodinium polyedra. Mar. Biol.,* **1996**, *127*, 35-44.

[41] Tilstone, G.H.; Airs, R.L.; Martinez-Vicente, V.; Widdicombe, C.; Llewellyn, C. High concentrations of mycosporine-like amino acids and colored dissolved organic matter in the sea surface microlayer off the Iberian Peninsula. *Limnol. Oceanogr.,* **2010**, *55*(5), 1835-1850.

CHAPTER 4

Analytical and Preparative Methods for MAAs

Abstract: This chapter describes the basics of analytical and preparative methods for mycosporine-like amino acids (MAAs). For samples whose molecular structures are known, high-performance liquid chromatography is widely used as a simple quantitative or qualitative analytical method for MAAs. However, if the molecular structures are unknown, they are often identified by combining several analytical methods, such as liquid chromatography-mass spectrometry and nuclear magnetic resonance analysis. In MAA preparation, the first key factor is how efficiently MAAs can be obtained in the extraction process from biological samples. The second key factor is how efficiently high-purity MAAs can be obtained from the separation process. This chapter also discusses the production of MAAs from an industrial perspective.

Keywords: Absorption spectrum, Amino acid analysis, Extraction, High-performance liquid chromatography, Industrial production, Liquid chromatography-mass spectrometry, Mass spectrometry, Nuclear magnetic resonance, Octadecyl silyl, Preparation, Purification, Reversed-phase chromatography.

INTRODUCTION

Analysis of Maas and their Molecular Structures

This section outlines various methods for the analysis of MAAs and methods for the identification of MAA molecules whose molecular structures are unknown.

HPLC Analysis of MAAs

When analyzing MAAs accumulated in organisms such as algae and cyanobacteria, samples extracted using a protonic solvent such as methanol or methanol are usually separated by HPLC instrument. MAAs are detected using wavelengths close to their absorption maximum. A technique called reversed-phase chromatography (RPC) is commonly used to analyze MAAs [1 - 3]. One of the column packing materials used in RPC is a silica gel to which alkyl groups are chemically bonded. The most widely used column is an ODS column (C18 column) filled with silica gel to which octa decyl silyl (ODS, $C_{18}H_{37}Si$) groups are

Hakuto Kageyama

bonded. Because octadecylsilyl groups show a small polarity, a substance with a smaller polarity and a higher hydrophobicity interacts more strongly with the ODS column and takes longer to elute from the column. By utilizing this property, the target substance contained in the mixed solution can be separated. For example, as shown in (Fig. **1**), in shinorine, mycosporine-2-glycine, and porphyra-334, the amino acid residues substituted at the C1 position of the cyclohexeneimine ring, which is the core structure of an MAA, are serine, glycine, and threonine, respectively.

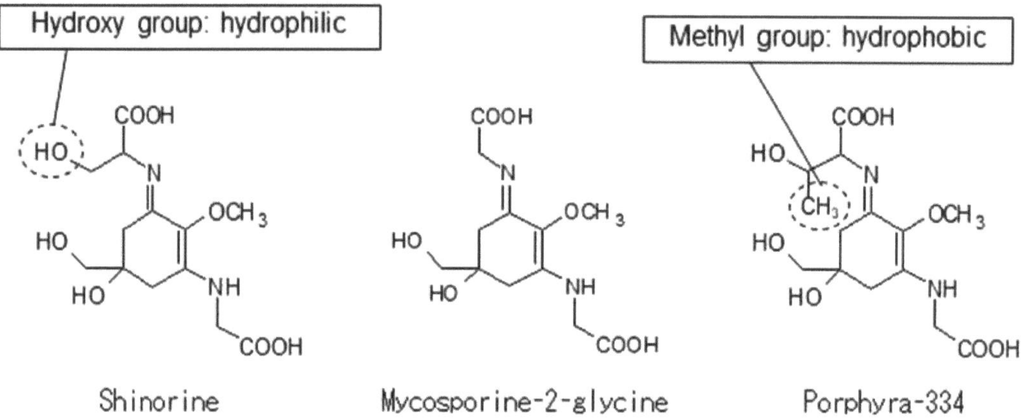

Fig. (1). Molecular structure of shinorine, mycosporine-2-glycine, and porphyra-334.

Fig. (**2**) shows the results of analyzing a mixed solution of these three MAAs using an ODS column. In this example, after injection of the sample into the column, shinorine is eluted first, followed by mycosporine-2-glycine, and finally porphyra-334. This result can be explained as follows. By comparing the molecular structures of shinorine and mycosporine-2-glycine, it can be seen that hydrophilic hydroxyl groups are present in the serine residue of shinorine (Fig. **1**). It is considered that the retention of shinorine in the column is weakened due to the influence of the presence of highly polar hydroxyl groups. However, the threonine residue of porphyra-334 also has a hydroxyl group like shinorine, but it is considered that the methyl group existing near it has a greater hydrophobic influence (Fig. **1**). Because the methyl group is hydrophobic, it is considered that its presence enhances the interaction between porphyra-334 and the column packing materials, resulting in a longer retention time. In the analysis of Fig. (**2**), 1% acetic acid aqueous solution was used as the mobile phase, but the separation pattern will differ depending on the solvent used as the mobile phase even if the same column is used. Although several methods have been proposed as general purpose methods for analyzing MAAs, it is important to find the optimum separation conditions according to the molecular structures of the target MAAs.

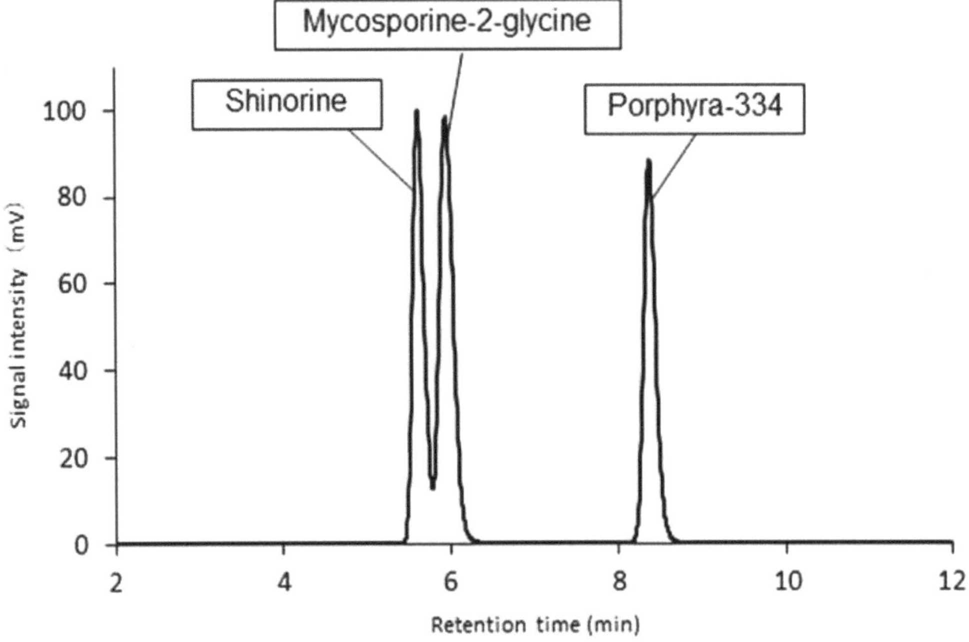

Fig. (2). HPLC chromatograph of a mixture that contained shinorine, mycosporine-2-glycine, and porphyra-334.

Methods to Determine the Molecular Structures of MAAs

When determining the types of MAAs by HPLC analysis, it is essential to confirm that the retention time of the analyte is consistent with that of the reference sample (authentic standard sample). In addition to matching the retention times, it is also necessary to confirm that the shape of the overlapping peak does not change when a mixture of the analyte and the reference sample is analyzed. This process is called spiking. When spiking is performed, if the analyte and the reference sample are different substances, changes in peak shape such as widening of the peak may be observed. After confirming these properties, if the retention time of the analyte and the reference sample match even if the analytical conditions, such as the composition of the mobile phase are changed, it is considered that these are the same substances. If the above analyses are performed and the retention times do not match, the same analyses should be repeated using other reference materials. Alternatively, the molecular structures of the analyte MAAs can be determined using analytical techniques other than HPLC. These analytical methods include measurement of absorption maximum wavelength, determination of molecular weight by mass spectrometry (MS), identification of amino acid substituents contained in the analyte MAA by amino acid analysis, and nuclear magnetic resonance (NMR) analysis.

Measurement of Absorption Maximum Wavelength of MAA

The absorption maximum wavelength, which varies depending on the molecular structure of MAAs, can be easily investigated using a spectrophotometer. However, some MAAs have very close values of the absorption maximum wavelength. Therefore, it can be difficult to estimate the structures of an analyte MAA using the absorption maximum wavelength alone. For example, as shown in (Fig. **3**), the absorption spectra of the purified products of shinorine, porphyra-334, and mycosporine-2-glycine are difficult to distinguish. In addition, as described in Chapter 1, the value of the absorption maximum wavelength may change depending on the type and pH of the dissolved solvent, so it is necessary to account for these conditions.

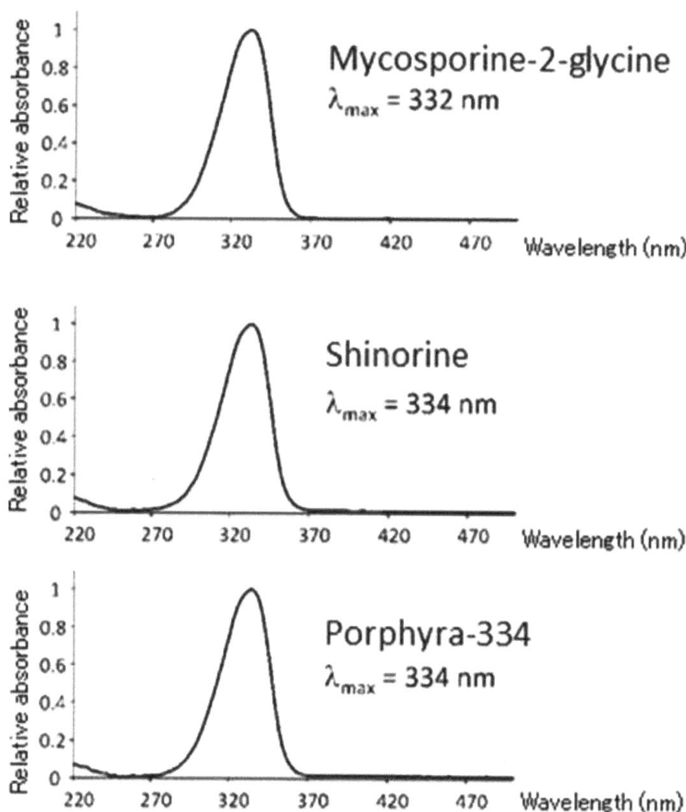

Fig. (3). Absorption spectra of Shinorine, porphyra-334, and mycosporine-2-glycine. These purified MAAs were dissolved in water and used for measurement.

Determination of Molecular Weight of MAA Molecules

Molecular weight information obtained by mass spectrometric analysis is important in the identification of MAA molecules. In the analysis of MAAs, liquid chromatography-mass spectrometry (LC/MS), which is a combination of HPLC and mass spectrometry, is the most widely used. (Fig. **4**) shows the result of the LC/MS analysis for 7-O-(β-arabinopyranosyl)-porphyra-334 (478-Da MAA) isolated from the cyanobacterium *Nostoc commune*. As described above, RPC is usually used as the analysis condition for LC to separate MAAs. An aqueous solution of a volatile acid such as formic acid or acetic acid is often used as the mobile phase.

HPLC condition

Mobile phase A: 0.1% formic acid (aqueous solution)
Mobile phase B: 0.1% formic acid (acetonitrile solution)
Column: InertSustain® AQ-C18, 1.9 µm, 2.1 mmID x 150 mm
Gradient program: 2%B, 0-5 min
95%B, 6-10 min
2%B, 10.5-25 min

MS condition

LTQ orbitrap discovery, ESI+
Source voltage (kV): 4.0
Sheath gas flow rate: 30.0
Aux gas flow rate: 10.0
Sweep gas flow rate: 0.0
Capillary temp (°C): 275
Capillary voltage (V): 10
Tube lens voltage (V): 80

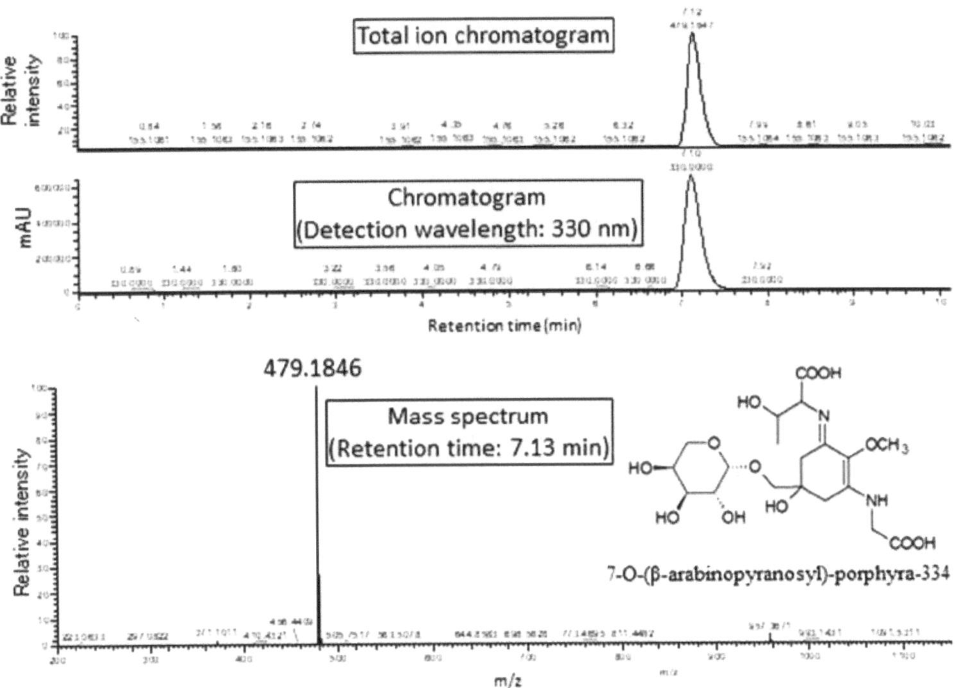

Fig. (4). LC/MS analysis of 7-O-(β-arabinopyranosyl)-porphyra-334.

Identification of Amino Acid Residues Contained in MAA Structures by Amino Acid Analysis

If the MAA molecule to be identified contains amino acid residues, the amino acid residues released by hydrolysis can be identified by amino acid analysis. Amino acid analysis is well established, and there are fully automated devices. (Fig. **5**) shows the results of the analysis of a standard sample of 17 amino acids using the Amino Acid Analyzer L-8900 (Hitachi High-Tech Corporation). There are several methods for hydrolysis treatment, such as using hydrochloric acid or mercapto ethane sulfonic acid. However, the experimental conditions need to be optimized because not all amino acids are recovered by each hydrolysis method.

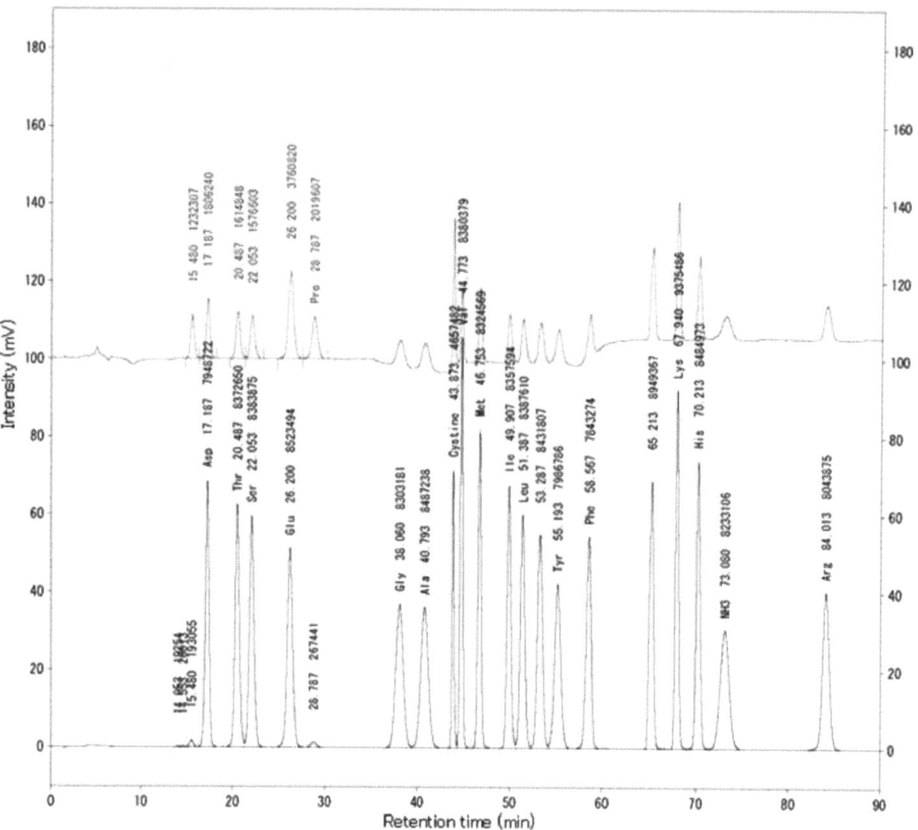

Fig. (5). Chromatogram of amino acid analysis using standard samples of 17 amino acids. This is the result of a post-column derivatization method using ninhydrin as a reaction reagent. The black and green chromatograms were detected at wavelengths of 570 nm and 440 nm, respectively. In this method, primary amines were detected at 570 nm, and secondary amines were detected at 440 nm. Cysteine was detected as cystine. The detected NH_3 was considered to be derived from the amide nitrogen in the protein, but it might also have been derived from air when preparing the sample solution.

Characterization of the Molecular Structure of MAA by NMR Analysis

The molecular structures of unknown MAAs are generally characterized by NMR analysis. In NMR analysis, a sample is placed in a strong magnetic field. The molecular structure of an organic compound is analyzed by utilizing the property that atomic nuclei will resonate at specific frequencies when a radiofrequency pulse is applied to the sample. The purity of the sample can also be quantified. Below are ^{1}H and ^{13}C NMR data of methanol-d_4 solutions of shinorine, poriphyra-334, and mycosporine-2-glycine as examples. In addition, Figs. (**6–8**) present the NMR spectra of each MAA [3].

Shinorine

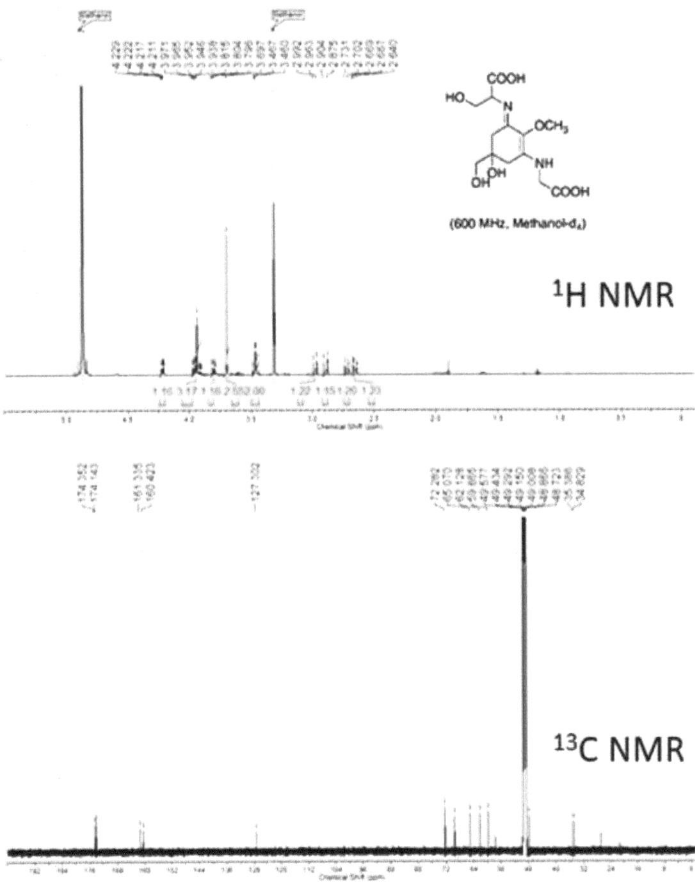

Fig. (6). NMR spectra of shinorine. The NMR data is shown on the next page.

Shinorine

^{1}H NMR (600 MHz, Methanol-d_4) δ ppm 4.22 (1H, dd, J = 7.1, 3.8 Hz), 3.90–3.98 (3H, m), 3.80 (1H, dd, J = 11.5, 7.1 Hz), 3.70 (3H, s), 3.48, 3.45 (2H, ABq, J = 11.4 Hz), 2.98 (1H, d, J = 17.2 Hz), 2.89 (1H, d, J = 17.2 Hz), 2.72 (1H, d, J = 17.2 Hz), 2.65 (1H, dd, J = 17.2, 0.9 Hz).

^{13}C NMR (151 MHz, Methanol-d_4) δ ppm 174.4, 174.1, 161.3, 160.4, 127.3, 72.3, 69.5, 65.1, 62.1, 59.9, 47.9, 35.4, 34.8.

HRMS (ESI-TOF^{+}) m/z: calcd. for $C_{13}H_{21}N_2O_8$ [M + H]$^{+}$: 333.1292, found: 333.1290.

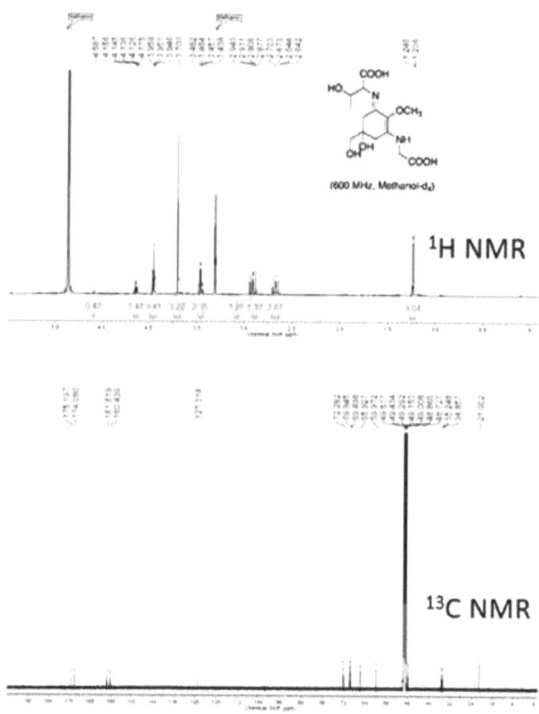

Fig. (7). NMR spectra of porphyra-334.

Porphyra-334

^{1}H NMR (600 MHz, Methanol-d_4) δ ppm 4.14 (1H, dt, J = 11.9, 6.1 Hz), 3.91-3.98 (3H, m), 3.70 (3H, s), 3.47, 3.45 (2H, ABq, J = 11.3 Hz), 2.93 (1H, d, J = 17.2 Hz), 2.89 (1H, d, J = 17.2 Hz), 2.67 (2H, m), 1.24 (3H, d, J = 6.4 Hz).

^{13}C NMR (151 MHz, Methanol-d_4) δ ppm 175.2, 174.1, 161.6, 160.4, 127.3, 72.3, 69.8, 69.5, 65.9, 60.0, 48.0, 35.2, 34.9, 21.0.

HRMS (ESI-TOF$^+$) *m/z*: calcd. for $C_{14}H_{23}N_2O_8$ $[M + H]^+$: 347.1449, found: 347.1449.

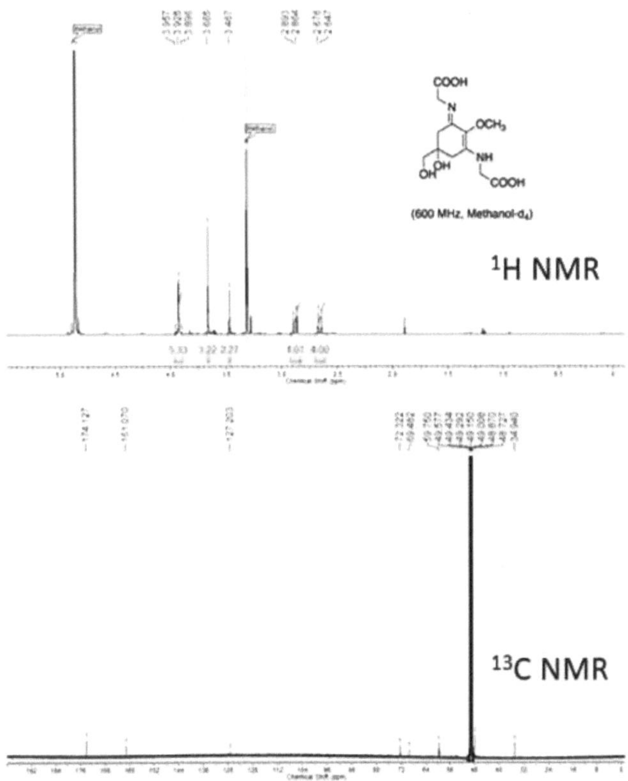

Fig. (8). NMR spectra of mycosporine-2-glycine.

Mycosporine-2-glycine

^1H NMR (600 MHz, Methanol-d_4) δ ppm 3.91, 3.93 (4H, ABq, $J = 17.4$ Hz), 3.67 (3H, s), 3.47 (2H, s), 2.88 (4H, d, $J = 17.4$ Hz), 2.66 (4H, d, $J = 17.4$ Hz).

^{13}C NMR (151 MHz, Methanol-d_4) δ ppm 174.1, 161.1, 127.2, 72.3, 69.5, 59.8, 47.9, 34.9.

HRMS (ESI-TOF$^+$) *m/z*: calcd. for $C_{12}H_{19}N_2O_7$ $[M + H]^+$: 303.1187, found: 303.1184.

NMR Conditions

Equipment: Avance III HD 600 NMR spectrometer equipped with a CryoProbe Prodigy (Bruker).

NMR tube: NMR test tube HG-type (Wako Pure Chemical Industries, Ltd.).

Reagents: Methanol-d_4, for NMR (Acros Organics).

The chemical shift values were based on the residual CD_2HOD (δH 3.31) and CD_3OD (δC 49.15). The ^{1}H-NMR spectrum is shown as δ (proton number, multiplicity, coupling constant J Hz). Multiplicity is indicated by s (singlet), d (doublet), and ABq (AB quartet).

Examples of Studies that Determined the Molecular Structures of MAAs

Below are some examples of studies that determined the molecular structures of MAAs. A research group at Meijo University attempted to identify an MAA obtained from the halotolerant cyanobacterium *Halothece* sp. PCC7418. First, they estimated that the target MAA was mycosporine-2-glycine by the *m/z* values of several signal peaks obtained by matrix-assisted laser desorption/ionization time-of-flight mass spectrometry (MALDI-TOF MS) analysis. Furthermore, when this MAA was hydrolyzed and analyzed for amino acids, only glycine was detected [4]. From these results, they concluded that the substance was mycosporine-2-glycine (later further confirmed by NMR analysis [3]). Mycosporine-2-glycine is an MAA with a typical molecular structure. Therefore, its molecular weight, molecular structure, and absorption maximum have been determined by analysis using other salt-tolerant cyanobacteria [5]. Mycosporine-2-glycine in the *Halothece* sp. PCC7418 strain was relatively easy to identify. However, to determine the molecular structures of novel MAAs, NMR analysis is indispensable. Moreover, it is not easy to determine the structures of MAAs that have complicated structures. For example, a research group at Kanazawa University identified an MAA isolated from the cyanobacterium *N. commune* by sequentially performing several analytical methods as follows. First, MALDI-TOF MS analysis determined that the molecular weight of this MAA was 478 Da. Next, it was estimated that this MAA was composed of porphyra-334 and pentose by infrared (IR) spectroscopy, MALDI-TOF MS/MS, and NMR analysis [6]. Finally, gas chromatography-mass spectrometry (GC/MS) analysis revealed that the pentose released by hydrolyzing this MAA was arabinose. Therefore, it was concluded that this MAA was 7-*O*-(β-arabinopyranosyl)-porphyra-334 [7].

PREPARATION AND PRODUCTION OF MAAS

In this section, MAA preparation methods including the extraction process and the chromatographic separation process are described with practical examples. The production of MAAs is also described from an industrial point of view.

Preparation of MAAs

In order to investigate the properties of each MAA, it is necessary to isolate and purify the target MAA from the organisms that accumulate MAAs. However, because MAAs have high water solubility, various water-soluble substances can be present as impurities. Therefore, it is not easy to obtain a high purity in separation processes that use an aqueous solution.

Extractions of MAAs

When extracting MAAs, first the microbial cells, algal tissues, etc. are chemically or physically disrupted in a solvent. It is common to use ultrasonic treatment or tissue crushing treatment with a mixer. Alternatively, cells, tissues, etc. are processed into a dry powder in advance, and a solvent is added to the powder for extraction. Because MAAs are easily dissolved in an aqueous solution, an aqueous solution of methanol or ethanol, which is a polar solvent, is frequently used. When extracting MAAs from cyanobacteria, methanol (99.5–99.8% by mass fraction for commercially available reagents) can be generally used [3]. However, for red algae, the optimal composition of the polar solvent for extraction of MAAs has been reported to differ depending on the strain, so it is necessary to optimize the conditions [8]. Temperature can also affect the extraction efficiency, and it is important to find optimal conditions that do not cause MAAs degradation.

Chromatographic Separation

After removing debris derived from unbroken cells or tissues by centrifugation, the target MAAs are often isolated and purified by preparative HPLC using a column with an inner diameter of 20 to 50 mm. Alternatively, low pressure liquid chromatography (LPLC) is a lower-cost option [3]. In most liquid chromatography, the operation is performed in RPC mode. Organic acids such as acetic acid and trifluoroacetic acid (TFA) are often added to the mobile phase in RPC mode. If these substances interfere with the intended use after purification, it is necessary to replace them with a suitable solvent using a technique such as gel filtration chromatography. In addition to the RPC mode, it is possible to separate MAAs using other separation modes such as ion exchange. Impurities can be removed by adsorbing MAAs on an activated carbon column [9] or by preparative thin-layer chromatography (TLC) [10].

Practical Example

Fig. (**9**) shows the procedure for separating mycosporine-2-glycine from the halotolerant cyanobacterium *Halothece* sp. PCC7418 [3].

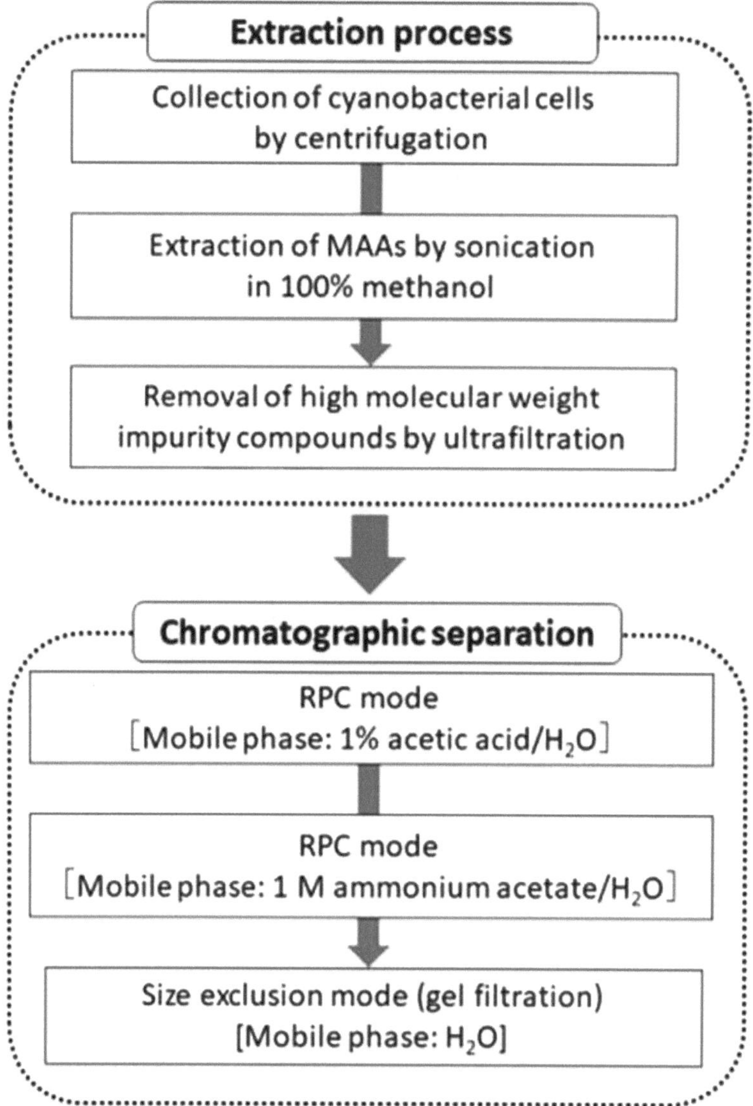

Fig. (9). Procedure for preparation of mycosporine-2-glycine from *Halothece* sp. PCC7418 cells.

Production of MAAs

Given their useful activities, MAAs have applications in the industrial field as cosmetics and pharmaceuticals. In fact, a number of MAA-related patents have been filed around the world [11]. Production strategy is important for the industrial application of MAAs. The use of fast-growing photosynthetic microorganisms such as cyanobacteria as a production platform is attractive from the perspective of sustainable production. Cyanobacteria can be cultivated in various systems such as field pools and closed culture systems constructed in the laboratory [12]. However, as far as the author knows, the construction of a cyanobacterial culture system for large-scale industrial production of MAAs has not been achieved so far. One of the reasons for the obstacles to the construction of mass production systems is the low MAA content of cyanobacteria. For example, the authors prepared mycosporine-2-glycine from the halotolerant cyanobacterium *Halothece* sp. PCC7418 strain. The cell culture contained only about 1 mg of mycosporine-2-glycine per liter. The yield of mycosporine--glycine after the purification process was only about 350 μg per liter.

Alternatively, as described in Chapter 3, one option is to introduce the cyanobacterial MAA biosynthetic genes into other host microbial cells to create transformed microorganisms capable of efficiently producing MAAs. *Escherichia coli* is often used as a host microorganism. When the authors optimized the culture conditions using *E. coli* in which the mycosporine-2-glycine biosynthesis gene of the halotolerant cyanobacterium *Halothece* sp. PCC7418 strain was introduced, the maximum amount of mycosporine-2-glycine was about 750 μg per liter of culture medium. Although this amount was lower than the *Halothece* sp. PCC7418 strain itself (1 mg/L), efficient production can be expected because the growth rate of *E. coli* is overwhelmingly faster than the *Halothce* sp. PCC7418 strain. There is room to consider the use of organisms other than *E. coli*. For example, 68.4 mg of shinorine was produced per liter of culture of the budding yeast *Saccharomyces cerevisiae* [13]. In this culture system, xylose for increasing the production of sedoheptulose-7-phosphate (S7P), which is a precursor of shinorine, and glucose for promoting growth were added to the culture medium. Furthermore, a strain lacking hexokinase, which catalyzes the phosphorylation reaction of glucose in glycolysis, was used. These strategies seem to have redirected the carbon flux from glycolysis into the pentose phosphate pathway, resulting in the efficient conversion of glucose to S7P (see Chapter 3 for more information on the biosynthetic pathways of MAAs). As another example, a method for obtaining MAAs from the extracellular medium of the microbial culture has been proposed. When the bacterium *Streptomyces avermitilis* was used as the host microorganism, shinorine, porphyra-334, and mycosporine-glycin--alanine were contained in the extracellular culture medium after culturing for 2 weeks at about 1400, 140, and 25 mg per liter, respectively [14]. Table **1** shows a collection of MAAs production report examples, including the cases mentioned above. Although there are limited reports showing the quantitative production of MAAs from microorganisms, it is found that *Streptomyces* may be suitable for MAAs production.

Table 1. Report examples of MAAs production.

Host Organism	Genes Introduced	Production of MAAs	Ref
Cyanobacterium *Halothece* sp. PCC7418 (wild type)	-	1.0 mg/L (mycosporine-2-glycine)	[3]
Escherichia coli	MAA biosynthetic genes of a cyanobacterium *Halothece* sp. PCC7418	750 µg/L (mycosporine-2-glycine)	[15]
Escherichia coli	MAA biosynthetic genes of a cyanobacterium *Nostoc linckia NIES-25*	3.5 mg/L (porphyra-334) 2.7 mg/L (palythine-threonine)	[16]
Saccharomyces cerevisiae	MAA biosynthetic genes of a cyanobacterium *Nostoc punctiforme*	68.4 mg/L (shinorine)	[13]
Streptomyces avermitilis	MAA biosynthetic genes of *Actinomycetales Actinosynnema mirum*	154 mg/L (shinorine)	[9]
Streptomyces avermitilis	MAA biosynthetic genes of *Actinomycetales Actinosynnema mirum*	1400 mg/L (shinorine) 140 mg/L (porphyra-334) 25 mg/L (mycosporine-glycine-alanine)	[14]

In the future, in order to further increase the production of MAAs using microorganisms, it would not be sufficient to simply introduce and overexpress MAA biosynthetic genes. Genetic manipulations that enhance the supply of methyl groups and specific amino acids required for MAAs synthesis might be effective. It might also be worth trying genetic manipulations to promote the biosynthesis of MAAs precursors. See Chapter 3 for details on the regulation of biosynthetic pathway of MAAs.

In addition, in order to achieve mass production, it is also necessary to develop an isolation and purification process for MAAs from a large amount of biological samples or culture medium. Besides separation by liquid chromatography, there is room for studying other separation methods such as adsorption, solid phase extraction (SPE), and crystallization. Recently, a separation method has been proposed that applied fast unsubstituted partition chromatography, which can separate samples using the principle of liquid–liquid extraction without using general columns [17].

CONCLUDING REMARKS

In this chapter, the analytical and preparative strategies for MAAs were summarized. Liquid chromatography is commonly used for the analysis of MAAs. However, various analytical methods are options to characterize the molecular structure of MAAs. From the viewpoint of mass production, the main sources of MAAs are algae and microbial cells. Liquid chromatography is often used in the MAA preparation process, but other methods have been studied and are expected to be developed in the future. The development of production processes using alternative approaches, such as chemical synthesis of MAAs, is also expected.

LIST OF ABBREVIATIONS

GC/MS	gas chromatography mass spectrometry
HPLC	high performance liquid chromatography
IR	infrared
LC/MS	liquid chromatography mass spectrometry
LPLC	low pressure liquid chromatography
MALDI-TOF MS	matrix-assisted laser desorption/ionization time-of-flight mass spectrometry
MAA	mycosporine-like amino acid
MS	mass spectrometry
NMR	nuclear magnetic resonance
ODS	octa decyl silyl
RPC	reversed-phase chromatography
SPE	solid phase extraction
S7P	sedoheptulose-7-phosphate
TFA	trifluoroacetic acid

CONSENT FOR PUBLICATION

Not applicable.

CONFLICT OF INTEREST

The author declares no conflict of interest, financial or otherwise.

ACKNOWLEDGEMENTS

Declared none.

REFERENCES

[1] Carreto, J.I.; Carignan, M.O.; Montoya, N.G. A high-resolution reverse-phase liquid chromatography method for the analysis of mycosporine-like amino acids (MAAs) in marine organisms. *Mar. Biol.,* **2005**, *146*(2), 237-252.
[http://dx.doi.org/10.1007/s00227-004-1447-y]

[2] Volkmann, M.; Gorbushina, A.A. A broadly applicable method for extraction and characterization of mycosporines and mycosporine-like amino acids of terrestrial, marine and freshwater origin. *FEMS Microbiol. Lett.,* **2006**, *255*(2), 286-295.
[http://dx.doi.org/10.1111/j.1574-6968.2006.00088.x] [PMID: 16448508]

[3] Ngoennet, S.; Nishikawa, Y.; Hibino, T.; Waditee-Sirisattha, R.; Kageyama, H. A method for the isolation and characterization of mycosporine-like amino acids from cyanobacteria. *Methods Protoc.,* **2018**, *1*(4), 46.
[http://dx.doi.org/10.3390/mps1040046] [PMID: 31164584]

[4] Waditee-Sirisattha, R.; Kageyama, H.; Sopun, W.; Tanaka, Y.; Takabe, T. Identification and upregulation of biosynthetic genes required for accumulation of Mycosporine-2-glycine under salt stress conditions in the halotolerant cyanobacterium *Aphanothece halophytica. Appl. Environ. Microbiol.,* **2014**, *80*(5), 1763-1769.
[http://dx.doi.org/10.1128/AEM.03729-13] [PMID: 24375141]

[5] Kedar, L.; Kashman, Y.; Oren, A. Mycosporine-2-glycine is the major mycosporine-like amino acid in a unicellular cyanobacterium (*Euhalothece* sp.) isolated from a gypsum crust in a hypersaline saltern pond. *FEMS Microbiol. Lett.,* **2002**, *208*(2), 233-237.
[http://dx.doi.org/10.1111/j.1574-6968.2002.tb11087.x] [PMID: 11959442]

[6] Matsui, K.; Nazifi, E.; Kunita, S.; Wada, N.; Matsugo, S.; Sakamoto, T. Novel glycosylated mycosporine-like amino acids with radical scavenging activity from the cyanobacterium *Nostoc commune. J. Photochem. Photobiol. B,* **2011**, *105*(1), 81-89.
[http://dx.doi.org/10.1016/j.jphotobiol.2011.07.003] [PMID: 21813286]

[7] Nazifi, E.; Wada, N.; Asano, T.; Nishiuchi, T.; Iwamuro, Y.; Chinaka, S.; Matsugo, S.; Sakamoto, T. Characterization of the chemical diversity of glycosylated mycosporine-like amino acids in the terrestrial cyanobacterium *Nostoc commune. J. Photochem. Photobiol. B,* **2015**, *142*, 154-168.
[http://dx.doi.org/10.1016/j.jphotobiol.2014.12.008] [PMID: 25543549]

[8] Sun, Y.; Han, X.; Hu, Z.; Cheng, T.; Tang, Q.; Wang, H.; Deng, X.; Han, X. Extraction, isolation and characterization of mycosporine-like amino acids from four species of red macroalgae. *Mar. Drugs,* **2021**, *19*(11), 615.
[http://dx.doi.org/10.3390/md19110615] [PMID: 34822486]

[9] Miyamoto, K.T.; Komatsu, M.; Ikeda, H. Discovery of gene cluster for mycosporine-like amino acid biosynthesis from *Actinomycetales* microorganisms and production of a novel mycosporine-like amino acid by heterologous expression. *Appl. Environ. Microbiol.,* **2014**, *80*(16), 5028-5036.
[http://dx.doi.org/10.1128/AEM.00727-14] [PMID: 24907338]

[10] Carreto, J.I.; Carignan, M.O.; Daleo, G.; Marco, S.G.D. Occurrence of mycosporine-like amino acids in the red-tide dinoflagellate *Alexandrium excavatum* : UV-photoprotective compounds? *J. Plankton Res.,* **1990**, *12*(5), 909-921.
[http://dx.doi.org/10.1093/plankt/12.5.909]

[11] Kageyama, H.; Waditee-Sirisattha, R. Mycosporine-like amino acids as multifunctional secondary metabolites in cyanobacteria: From biochemical to application aspects. In: *Studies in Natural Products Chemistry*; Atta-ur-Rahman, , Ed.; Elsevier, **2018**; 59, pp. 153-194.

[12] Grewe, C.B.; Pulz, O. The biotechnology of cyanobacteria. In: *Ecology of Cyanobacteria II*; , **2012**; pp. 707-739.
[http://dx.doi.org/10.1007/978-94-007-3855-3_26]

[13] Jin, C.; Kim, S.; Moon, S.; Jin, H.; Hahn, J.S. Efficient production of shinorine, a natural sunscreen material, from glucose and xylose by deleting *HXK2* encoding hexokinase in *Saccharomyces cerevisiae*. *FEMS Yeast Res.,* **2021**, *21*(7), foab053.
[http://dx.doi.org/10.1093/femsyr/foab053] [PMID: 34612490]

[14] Method for producing mycosporin-like amino acids using microorganisms. Kitasato Institute; Nagase Co Ltd., **2015**.

[15] Patipong, T.; Hibino, T.; Waditee-Sirisattha, R.; Kageyama, H. Efficient bioproduction of mycosporine-2-glycine, which functions as potential osmoprotectant, using *Escherichia coli* cells. *Nat. Prod. Commun.,* **2017**, *12*(10), 1934578X1701201.
[http://dx.doi.org/10.1177/1934578X1701201017]

[16] Chen, M.; Rubin, G.M.; Jiang, G.; Raad, Z.; Ding, Y. Biosynthesis and heterologous production of mycosporine-like amino acid palythines. *J. Org. Chem.,* **2021**, *86*(16), 11160-11168.
[http://dx.doi.org/10.1021/acs.joc.1c00368] [PMID: 34006097]

[17] Zwerger, M.; Schwaiger, S.; Ganzera, M. Efficient isolation of mycosporine-like amino acids from marine red algae by fast centrifugal partition chromatography. *Mar. Drugs,* **2022**, *20*(2), 106.
[http://dx.doi.org/10.3390/md20020106] [PMID: 35200636]

CHAPTER 5

Biological Activities of MAAs and their Applications 1: UV-protective Activity of MAAs and their Application as Sunscreens

Abstract: In this chapter, we focus on the UV absorption characteristics of MAAs and describe the application examples. UV rays that pass through the ozone layer and the atmosphere and reach the surface of the Earth consist of UV-A and UV-B. Because these rays are harmful to biomolecules, MAAs, which can efficiently absorb these wavelength regions and detoxify their by-products, are promising natural organic compounds such as sunscreens. Products containing MAAs extracted from red algae are already on the market. With a focus on Helioguard 365 and HELINORI, the biological effects of MAAs against UV irradiation will be described.

Keywords: Collagen, DNA damage, Elastin, Helioguard 365, HELINORI, Melanin, Melanocyte, Photoaging, *Porphyra umbilicalis*, Red alga, UV-A, UV-B, UV-C.

INTRODUCTION

The skin is the largest human organ and is constantly exposed to the surrounding environment. Among various environmental stresses, UV rays are known to have a negative effect on the skin. Exposure to UV light damages the skin, and its reaction mechanism depends on the wavelength of light.

UV rays are divided into three wavelength regions: UV-A, UV-B, and UV-C [1]. Of the UV rays contained in the Sun's rays, UV-A (315–400 nm) and UV-B (280–315 nm) reach the surface of the Earth without being absorbed by the ozone layer and the atmosphere (Fig. **1**). Among them, UV-A accounts for 95% of the total UV radiation. Although the proportion of UV-B is small, it is high energy and is considered to be more harmful to the skin and eyes than UV-A [2]. In recent decades, the amount of UV-B rays reaching the surface of the Earth has increased due to the depletion of the ozone layer [3, 4]. Both UV-A and UV-B are known to be genotoxic and cause photochemical damage to biopolymer compounds such as intracellular DNA and proteins [5, 6]. As a result, these irradiations accelerate skin aging and can lead to skin cancer. In contrast, although

Hakuto Kageyama

UV-C (100–280 nm) is high energy, it does not affect living organisms because it is absorbed by the ozone layer and the atmosphere and does not reach the surface of the Earth (Fig. **1**) [2, 7].

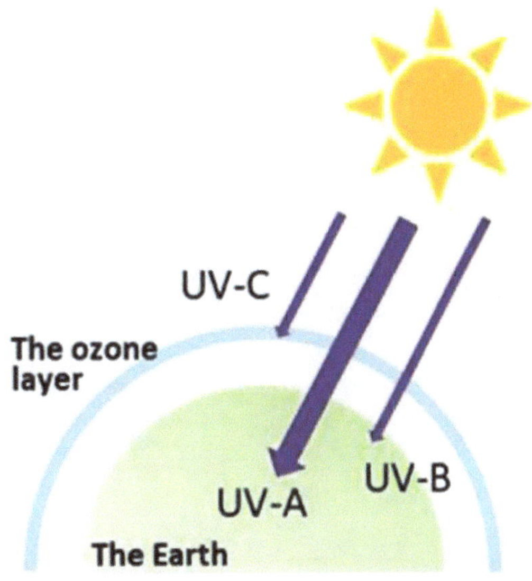

Fig. (1). UV irradiation that reaches the surface of the earth. 95% of the UV irradiation that reaches the surface of the Earth is UV-A. Some UV-B also reach the surface of the Earth. UV-C is completely absorbed by the ozone layer and the atmosphere.

UV-A

UV-A has a weaker energy than UV-B, but it has a large effect on the skin because the amount of radiation that reaches the surface of the Earth is large. UV-A irradiated to the skin reaches the dermis and interferes with the functions of collagen and elastin, which give the skin its firmness and elasticity (Fig. **2**). It also damages fibroblasts that produce collagen and elastin. As a result, the skin becomes less firm and elastic, causing wrinkles and sagging. In addition, by activation of melanocytes, it promotes the biosynthesis of melanin pigments and creates age spots. These actions are called photoaging.

Characteristics of UV-A and its Effect on the Skin

- UV-A is 95% of the UV rays that reach the surface of the Earth.
- Although its energy is weak, long-term exposure causes chronic damage.
- The wavelength is long and reaches the dermis.
- Causes wrinkles, sagging, and age spots.

UV-B

UV-B is only about 5% of the UV rays that reach the surface of the Earth, but it has higher energy. Because the depth of the reach of UV-B is shallow, it reaches only the epidermis (Fig. **2**). Because it damages the DNA in epidermal cells, the risk of developing skin cancer increases with long-term exposure*. Even short exposures can cause an inflammation reaction, resulting in redness of the skin. This is called sunburn. UV-B also causes suntan, which causes the skin to brown due to the deposition of melanin pigments several days after exposure. It can also cause age spots and freckles.

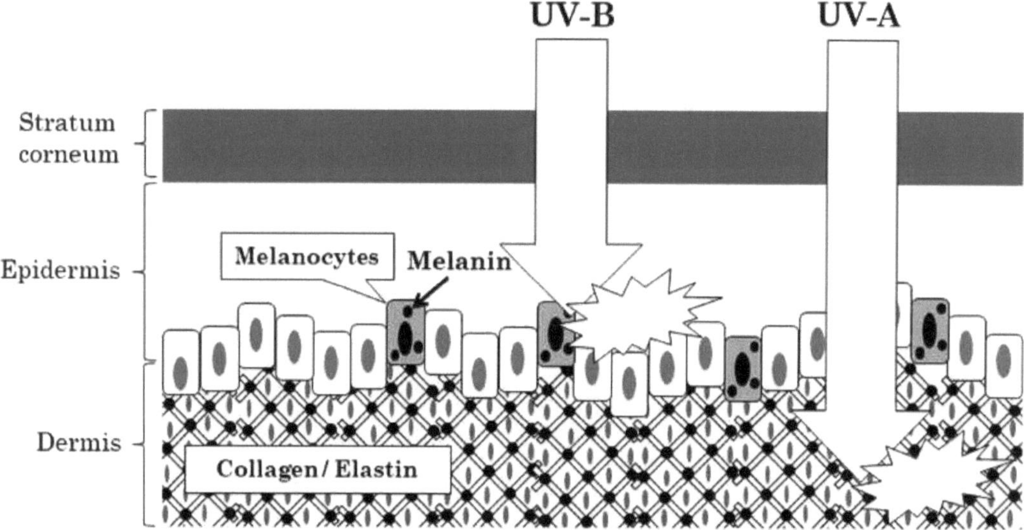

Fig. (2). Effects of UV-A and UV-B on the skin.

* UV-A is also known to indirectly damage DNA. UV-A irradiation produces reactive oxygen species (ROS) in the body, which causes oxidative damage to biopolymer compounds such as DNA [8].

Characteristics of UV-B and its Effect on the Skin

- It has higher energy than UV-A and causes DNA damage.
- The wavelength is short, and the reach of depth is shallow.
- Short-term exposure causes acute sunburn and suntan.

MAAs have an absorption maximum of 310 ~ 362 nm, which is in the range including UV-A and UV-B, and have large molar extinction coefficients (ε =

28,100–50,000 M^{-1} cm^{-1}) [9]. In addition, it is possible to convert absorbed UV into harmless thermal energy without generating ROS. Therefore, MAAs can be used as natural sunscreens.

Sunscreens containing MAAs are already on the market, and examples of well-known products are Helioguard 365 (Fig. **3**) from Mibelle Biochemistry (https://mibellebiochemistry.com/) in Switzerland and HELINORI from Biosil Technologies (http://www.biosiltech.com/) in the United States. These materials contain MAAs derived from the red alga *Porphyra umbilicalis* as ingredients. In some cases, cyanobacteria-derived MAAs are also used in skin care products. For example, Skinxia (https://www.instagram.com/skinxia_official/) and Lekarka (https://lekarka.co.jp/) in Japan sell skin care cosmetics containing MAAs extracted from cyanobacteria. Doctor's Choice (https://d-choice.net/), headquartered in the United States, has succeeded in extracting MAAs at a high concentration and is developing an original equipment manufacturing (OEM) business for cosmetics.

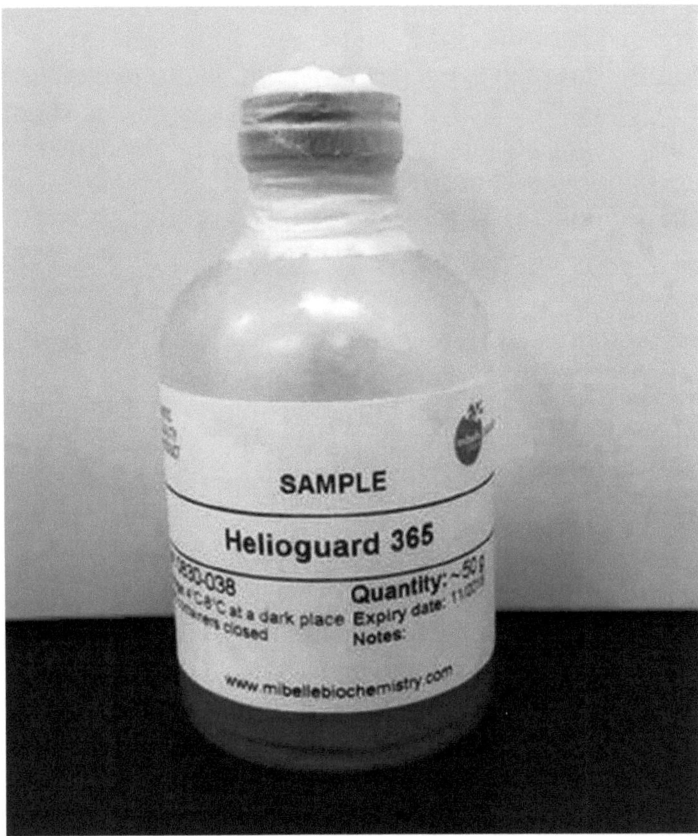

Fig. (3). Helioguard 365 sample.

HELIOGUARD 365

About Helioguard 365

Helioguard 365 contains shinorine and porphyra-334 extracted from the red alga *Porphyra umbilicalis*. These MAAs are liposomalized. The published preparation method for Helioguard 365 is as follows [10].

1. Add dried *P. umbilicalis* to a 15% aqueous ethanol solution to a concentration of 3.3%, and extract MAAs at 45°C for 2 h under constant stirring conditions.
2. After removing the debris of red alga, perform ultrafiltration with a 10 kDa cut-off membrane to obtain a transparent, MAA-containing extract.
3. Mix the MAA-containing extract with 3.3% lecithin to form liposomes, and add phenoxyethanol to a concentration of 0.4%.
4. Adjust the concentration of MAAs to 0.1%.

According to Mibelle Biochemistry*, the recommended concentration of Helioguard 365 when used as a sunscreen is 1–5% (MAA concentration 0.001–0.005%). Because the extract of *P. umbilicalis* may contain electrolytes, a gelling agent or emulsifier that is stable to the electrolytes should be selected when formulating. When ethanol is used, its concentration should not exceed 20% to prevent the liposomes from becoming unstable. Treatment below 50°C is desirable in the formulating process.

* https://mibellebiochemistry.com/how-formulate-helioguardtm-365

Properties and Activities of Helioguard 365

Helioguard 365 is suitable for daily skin care applications and can be expected to have an inhibitory effect on skin aging. Next, the activities of Helioguard 365 reported by Mibelle Biochemistry are described.

Stability of MAAs in Helioguard 365

When the stability of MAAs contained in Helioguard 365 was verified at 4°C, room temperature, and 37°C, no decrease in the amount of MAAs was confirmed after one month. After three months, a 20% reduction in MAAs was seen in samples kept at 37°C, but not at 4°C or room temperature. In samples subjected to UV-A irradiation treatment during heat insulation, the amount of MAAs did not decrease and existed stably at all temperatures even after 3 months [10].

Inhibitory Effect of Helioguard 365 on DNA Damage caused by UV Irradiation in Human Fibroblast Cells

UV irradiation causes DNA damage in human fibroblast cells. However, the damage to DNA was reduced by adding 3 to 5% Helioguard 365 to fibroblast cells before UV irradiation. These results were obtained by the comet assay, which is a simple method for measuring DNA strand breaks in cells [11].

Improvement of Skin Firmness, Smoothness, and Wrinkles by Helioguard 365

The following experimental results have been reported. A cream* prepared to contain 5% Helioguard 365 (final concentration of MAAs of 0.005%) was applied to the medial forearm and face of 20 women aged 36 to 54 years. They were treated with UV-A irradiation at 10 J/cm^2 twice a week, and then the elasticity, roughness, and depth of wrinkles of the skin were quantified. As a result, it was confirmed that the cream containing Helioguard 365 improved elasticity and smoothness and reduced the depth of wrinkles compared to the control cream not containing Helioguard 365 [10].

* The method of preparing the cream and the raw materials are not described in the cited references and are unknown.

Inhibitory Effect of Helioguard 365 on Lipid Peroxidation

Similar to the section above, applying a cream containing 5% Helioguard 365 was found to have an inhibitory effect on the skin's lipid peroxidation caused by UV-A irradiation [10]. The lipid peroxidation reaction is triggered by ROS, converting lipids into lipid peroxides. This reaction causes rough skin and wrinkles.

Enhancement Effect of Helioguard 365 on SPF Value

It has been confirmed that the value of the sun protection factor (SPF), which is an indicator of the UV protection effect of sunscreen agents, is increased by the addition of Helioguard 365. It was found that the addition of Helioguard 365 to a sunscreen agent of SPF7.2 to a concentration of 2% increased the SPF value to about 8.3 [11].

The larger the SPF value, the higher the effect of UV-B protection. If the value exceeds 50, SPF is expressed as 50+. Similarly, there is PA (protection grade of UV-A) as an index showing the degree of protection against UV-A irradiation. PA is divided into four stages of PA +, PA ++, PA +++, PA ++++, and the larger the number of +, the higher the UV-A protection effect.

HELINORI

About HELINORI

HELINORI contains palythine in addition to shinorine and porphyra-334, which are MAAs extracted from the red alga *P. umbilicalis* and used in Helioguard 365. The 'nori' part of HELINORI is derived from the Japanese language. (*P. umbilicalis* is edible as "nori" in Japan.).

Properties and Activities of HELINORI

The activity of HELINORI has been reported as follows.

Stability of Helonori

HELINORI was resistant to 6 h of sunlight exposure or 30 min of 120°C treatment and was stable for at least 18 months under temperature conditions of 15–25°C [12].

Suppression of Sunburn

HELINORI seems to give a similar sun protection as Helioguard 365. It has been reported that the application of a cream containing 5% HELINORI suppressed the formation of sunburn cells generated after UV irradiation [12].

Other Activities of Helonori

It was found that the membrane lipids of fibroblast cells and keratinocytes exposed to oxidative stress derived from UV-A irradiation were protected and biological activity was maintained in the presence of 2% HELINORI [12].

CONCLUDING REMARKS

MAAs are promising, natural UV-absorbing organic compounds. This chapter focused on the already commercialized Helioguard 365 and HELINORI and outlined their properties. It has been clarified that these formulations can reduce DNA damage and suppress photoaging of the skin against UV irradiation treatment. To date, various useful activities of MAAs have been reported in addition to the UV-absorbing property. These activities will be summarized in Chapters 6–10.

LIST OF ABBREVIATIONS

MAA mycosporine-like amino acid

ROS reactive oxygen species

UV ultraviolet

CONSENT FOR PUBLICATION

Not applicable.

CONFLICT OF INTEREST

The author declares no conflict of interest, financial or otherwise.

ACKNOWLEDGEMENTS

Declared none.

REFERENCES

[1] Gao, Q.; Garcia-Pichel, F. Microbial ultraviolet sunscreens. *Nat. Rev. Microbiol.*, **2011**, *9*(11), 791-802.
[PMID: 21963801]

[2] Gruber, F.; Peharda, V.; Kastelan, M.; Brajac, I. Occupational skin diseases caused by UV radiation. *Acta Dermatovenerol. Croat.*, **2007**, *15*(3), 191-198.
[PMID: 17868545]

[3] Rastogi, R.P.; Incharoensakdi, A. Analysis of UV-absorbing photoprotectant mycosporine-like amino acid (MAA) in the cyanobacterium *Arthrospira* sp. CU2556. *Photochem. Photobiol. Sci.*, **2014**, *13*(7), 1016-1024.
[PMID: 24769912]

[4] Nguyen, K.H.; Chollet-Krugler, M.; Gouault, N.; Tomasi, S. UV-protectant metabolites from lichens and their symbiotic partners. *Nat. Prod. Rep.*, **2013**, *30*(12), 1490-1508.
[PMID: 24170172]

[5] Ikehata, H. Mechanistic considerations on the wavelength-dependent variations of UVR genotoxicity and mutagenesis in skin: The discrimination of UVA-signature from UV-signature mutation. *Photochem. Photobiol. Sci.*, **2018**, *17*(12), 1861-1871.
[PMID: 29850669]

[6] Brown, N.; Donovan, F.; Murray, P.; Saha, S. Cyanobacteria as bio-factories for production of UV-screening compounds. *OA Biotechnology.*, **2014**, *3*, 6.

[7] Jallad, K.N. Chemical characterization of sunscreens composition and its related potential adverse health effects. *J. Cosmet. Dermatol.*, **2017**, *16*(3), 353-357.
[PMID: 27596093]

[8] Panich, U.; Sittithumcharee, G.; Rathviboon, N.; Jirawatnotai, S. Ultraviolet radiation-induced skin aging: The role of DNA damage and oxidative stress in epidermal stem cell damage mediated skin aging. *Stem Cells Int.*, **2016**, *2016*, 7370642.
[PMID: 27148370]

[9] Kageyama, H.; Waditee-Sirisattha, R. Antioxidative, anti-inflammatory, and anti-aging properties of mycosporine-like amino acids: Molecular and cellular mechanisms in the protection of skin-aging.

Mar. Drugs, **2019**, *17*(4), 222.
[PMID: 31013795]

[10] Schmid, D.; Schürch, C.; Zülli, F. Mycosporine-like amino acids from red algae protect against premature skin-sging. *Euro Cosmetics,* **2006**, 1-4.

[11] *The brochure of Helioguard 365 provided by H. Holstein Co., Ltd, Tokyo, Japan*; , **2006**.

[12] Navarro, N.; Figueroa, F.L.; Korbee, N.; Bonomi, J.; Gomez, F.A.; de la Coba, F. Mycosporine-like amino acids from red algae to develop natural UV sunscreens. In: *Sunscreens: Source, Formulations, Efficacy and Recommendations*; Rastogi, R.P., Ed.; Nova Science Publishers, **2018**.

CHAPTER 6

Biological Activities of MAAs and their Applications 2: Antioxidative Properties

Abstract: It is known that the generation of reactive oxygen species (ROS) caused by UV irradiation and oxidative reactions accelerate skin aging. Substances that suppress or eliminate the generation of ROS are called antioxidants. So far, various mycosporine-like amino acids (MAAs) have been reported to have antioxidative activities. To prevent damage to the skin caused by ROS and maintain the homeostasis of the epidermis, skin cells have an endogenous antioxidant system consisting of enzymatic reactions. Although many points are unclear about the regulatory mechanisms, it has been suggested that MAAs are involved in the regulation of genes encoding enzymes that are involved in this system. This chapter provides a comprehensive overview of the antioxidant activities of MAAs.

Keywords: Antioxidant, Catalase, Glutathione peroxidase, Glutathione reductase, Peroxiredoxin, Oxidation, Reactive oxygen species, Superoxide dismutase, Thioredoxin reductase.

INTRODUCTION

Reactive oxygen species (ROS) that trigger oxidative stress include the hydroxyl radical (•OH), superoxide anion radical ($\bullet O_2^-$), hydrogen peroxide (H_2O_2), and singlet state molecular oxygen ($^1O_2$). Within the skin, exposure to UV irradiation is associated with the production of ROS. The ROS generation reaction by UV irradiation is diverse and depends on its wavelength. For example, the production of $^1O_2$ and $\bullet O_2^-$ was promoted in the skin of UV-A irradiated mice [1]. There are also reports that • OH, $\bullet O_2^-$, and H_2O_2 were generated from advanced glycation end products (AGE) during UV-A irradiation [2]. Although the mechanism of formation is unknown, it has been reported that UV-B irradiation also induced • OH, $\bullet O_2^-$, and H_2O_2 [3].

Oxidation is an essential reaction in energy production and metabolic pathways, and these reaction processes are responsible for the production of ROS. The generated ROS function as signal transduction molecules that cause cell division, inflammation, immune function, stress response, *etc* [4]. In addition, in plants and

Hakuto Kageyama

photosynthetic microorganisms such as cyanobacteria, excess light energy is absorbed by the photosynthetic reaction system, and if this energy is not safely dispersed, ROS such as $^1O_2$ and H_2O_2 are generated.

Skin cells have an endogenous antioxidant system to prevent damage to the skin by ROS caused by UV irradiation and oxidative reactions and to regulate epidermal homeostasis. The system consists of six enzymes: superoxide dismutase (SOD), catalase (CAT), glutathione peroxidase (GPX), glutathione reductase (GR), thioredoxin reductase (TRXR), and peroxiredoxin (PRDX). SOD and CAT erase O_2^- and H_2O_2, respectively, and convert them into H_2O (Figs. **1** & **2**). In contrast, GPX, GR, TRXR, and PRDX eliminate H_2O_2 by regulating the redox state of glutathione and thioredoxin (Fig. **3**). In addition to this enzymatic system, non-enzymatic molecules such as ascorbic acid (vitamin C), α-tocopherol (vitamin E), and glutathione play important roles as antioxidants in the skin tissue [5]. These small molecule compounds eliminate free radicals by acting as electron donors (Fig. **4**).

$$2{\cdot}O_2^- + 2H^+ \xrightarrow{\text{SOD}} H_2O_2 + O_2$$

Fig. (1). The reaction catalyzed by SOD.

$$2H_2O_2 \xrightarrow{\text{CAT}} 2H_2O + O_2$$

Fig. (2). The reaction catalyzed by CAT.

ANTIOXIDATIVE ACTIVITIES OF MAAS

To date, many MAAs have been reported to exhibit antioxidative activites (Table **1**). Exposure to UV radiation causes the production of ROS, one of the factors that promote skin aging. Therefore, antioxidant molecules with ROS-scavenging activity are commonly used as cosmetic ingredients to prevent aging. Mycosporine-glycine is considered to be one of the molecular structures of MAAs with the strongest antioxidant activity. It has been reported that the antioxidant activity of mycosporine-glycine isolated from the lichen *Lichina pygmaea* is higher than that of shinorine and porphyra-334 under the condition of pH 8.5 by the ABTS assay. The half-maximal (50%) inhibitory concentration (IC_{50})* of mycosporine-glycine was 3 μM, which was significantly lower than the well-known antioxidant ascorbic acid (26 μM) [6]. Mycosporine-2-glycine has also been reported to have a higher antioxidant activity than shinorine or porphyra-334 [7]. However, the following points should be noted when discussing the antioxidative capacity of each MAA molecule. As can be seen from Table **1**, the evaluation of the antioxidant activity of MAAs differs depending on the measurement method and the research group. For example, three studies investigated the DPPH-free radical scavenging activity of porphyra-334. Of these reports, the activity of porphyra-334 was detected in two cases, but the IC_{50} values differed significantly between them [8 - 10]. One of the reasons for these discrepancies is considered to be the purity of the MAAs used in the experiments. As mentioned in Chapter 4, it is generally not easy to purify MAAs with high purity, so it is possible that impurities may have affected the measurements. There may also have been subtle differences in the measuring methods and the type and condition of the measuring instruments.

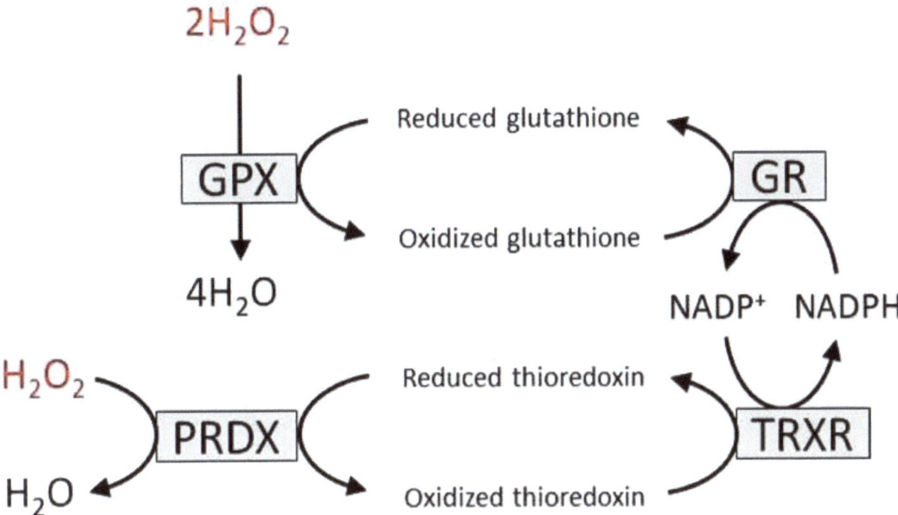

Fig. (3). The reactions catalyzed by GPX, GR, TRXR, and PRDX.

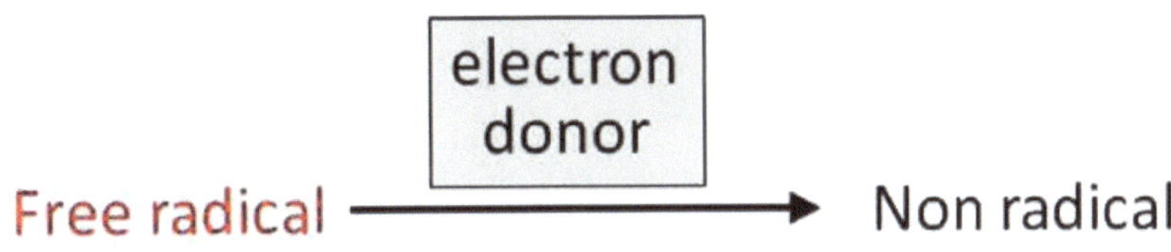

Fig. (4). A nonenzymatic reaction to neutralize free radicals by donating electrons from electron donors.

* The IC_{50} value is an index of the concentration at which half (50%) of the action of the target substance is inhibited. The lower the value, the more efficient the inhibition.

Table 1. Antioxidative activities of MAAs.

Method	Activities
Mycosporine-glycine	
PC-assay [11]	+
β-carotene bleaching method [6]	+
ABTS assay [6]	+ (IC_{50}: 20 μM (pH 6.0), 4 μM (pH 7.5), 3 μM (pH 8.5))
DPPH assay [8]	+ (IC_{50}: 4.2 μM)
DPPH assay [9]	+ (IC_{50}: 43 μM)
Superoxide assay [6]	-
Singlet oxygen quenching [12]	+ (Rate constant: 5.6×10^{7} M^{-1} s^{-1})
Mycosporine-alanine	
DPPH assay [13]	+ (IC_{50}: 7.6 mM)
Mycosporine-GABA	
ABTS assay [14]	+ (IC_{50}: 600 μM)
Shinorine	
PC-assay [11]	-
β-carotene bleaching method [6]	+
ABTS assay [6]	+ (IC_{50}: ND (pH 6.0), ND (pH 7.5), 100 μM (pH 8.5))
ABTS assay [7]	+ (IC_{50}: 94 μM)
DPPH assay [8]	-
DPPH assay [10]	+ (IC_{50}: 399 μM)
Superoxide assay [6]	+
Singlet oxygen quenching [15]	-
Porphyra-334	
PC-assay [11]	-
β-carotene bleaching method [6]	+
ABTS assay [6]	+ (IC_{50}: 1000 μM (pH 6.0), 400 μM (pH 7.5), 80 μM (pH 8.5))
ABTS assay [7]	+ (IC_{50}: 133 μM)
ABTS assay [16]	+ (IC_{50}: >72 μM (pH 5.8), >72 μM (pH 6.6), 28 μM (pH 7.4), 21μM (pH 8.0))
DPPH assay [8]	-
DPPH assay [9]	+ (IC_{50}: 3.4 mM)

(Table 1) cont.....

Method	Activities
DPPH assay [10]	+ (IC_{50}: 185 μM)
Superoxide assay [6]	+
Singlet oxygen quenching [15]	-
Mycosporine-2-glycine	
ABTS assay [7]	+ (IC_{50}: 40 μM)
DPPH assay [9]	+ (IC_{50}: 22 μM)
Asterina-330	
PC-assay [11]	-
β-carotene bleaching method [6]	+
ABTS assay [6]	+ (IC_{50}: 1000 μM (pH 6.0), 60 μM (pH 7.5), 10 μM (pH 8.5))
Superoxide assay [6]	+
Singlet oxygen quenching [15]	-
Palythine	
PC-assay [11]	-
ABTS assay [16]	+ (IC_{50}: >72 μM (pH 5.8), >72 μM (pH 6.6), 23 μM (pH 7.4), 12 μM (pH 8.0))
DPPH assay [10]	+ (IC_{50}: 21 μM)
ORAC assay [17]	+ (IC_{50}: 714 μM)
Singlet oxygen quenching [15]	-
Palythinol	
PC-assay [11]	-
Palythene	
Singlet oxygen quenching [15]	-
Usujirene	
FTC assay [18]	+
TBA assay [18]	+
Palythenic acid	
Singlet oxygen quenching [15]	-
Mycosporine-glutaminol-glucoside	
Singlet oxygen quenching [19]	+ (Rate constant: $5.9 \times 10^{7}\ M^{-1}\ s^{-1}$)
478-Da MAA (7-O-(β-arabinopyranosyl)-porphyra-334)	
DPPH assay [20]	-
ABTS assay [14]	+ (IC_{50}: 9.5 mM)
508-Da MAA (Hexose-bound porphyra-334)	

(Table 1) cont.....

Method	Activities
ABTS assay [14]	+ (IC_{50}: 58 mM)
612-Da MAA (Two hexose-bound palythine-threonine derivative)	
ABTS assay [14]	+ (IC_{50}: 16 mM)
Nostoc-756	
ABTS assay [21]	+ (IC_{50}: 515 μM)
880-Da MAA ({Mycosporine-ornithine: 4-deoxygadusol ornithine}-β-xylopyranosy--β-galactopyranoside)	
ABTS assay [14]	+ (IC_{50}: 510 μM)
1050-Da MAA (Mycosporine-2-(4-deoxygadusol-ornithine)-β-xylopyranosyl-β-galactopyranoside)	
DPPH assay [20]	+ (IC_{50}: 809 μM)
ABTS assay [14]	+ (IC_{50}: 1.0 mM)
ABTS assay [21]	+ (IC_{50}: 144 μM)
13-*O*-(β-galactosyl)-porphyra-334	
ABTS assay [22]	+ (IC_{50}: 17 mM)

+: Activity was detected.
-: Activity was not detected.

EFFECTS OF MAAS ON THE ANTIOXIDANT SYSTEM

ROS play a major role in aging because of their irreversible oxidative damage to biomolecules. To protect biomolecules from ROS, it is known that the endogenous enzymatic defense system composed of the reactions shown in Figs. (**1**–**3**) removes ROS. Interestingly, it has been reported that MAAs influenced the expression of antioxidant enzymes (SOD, CAT, GPX, GR, TRDX, PRDX), which are involved in the reactions (Table **2**).

First, an example with mycosporine-2-glycine will be given. In the RAW 264.7 macrophage cell line, the expression of *cat* and *sod1* genes induced by oxidative stress was suppressed by the addition of mycosporine-2-glycine purified from the halotolerant cyanobacterium *Halothece* sp. PCC7418 to the culture medium [23]. This phenomenon may have been caused by the free radical scavenging activity of mycosporine-2-glycine. In other words, it is possible that mycosporine-2-glycine reduced the intracellular oxidative stress, which suppressed the expression of the *cat* and *sod1* genes that were originally promoted under oxidative stress conditions. However, it was reported that when mycosporine-2-glycine accumulated intracellularly in the freshwater cyanobacteria *Synechococcus elongatus* PCC7942 by the introduction of biosynthetic genes derived from the halotolerant cyanobacterium, antioxidant genes including *cat* and *sodB* were upregulated under oxidative stress conditions [24]. It is interesting that the regulation of gene expression differed between macrophage cells and cyanobacterial cells. Mycosporine-2-glycine has also been reported to suppress oxidative stress-induced DNA damage in human malignant melanoma A375 cells and normal human skin fibroblasts CRL-1474 cells [9].

Studies investigating the effects of MAAs on antioxidant enzymes in animal tissues have also been reported. By applying an aqueous solution containing shinorine and porphyra-334 isolated from the red alga *Porphyra yezoensis* to mice, the decrease in antioxidant enzyme activity in back skin tissue caused by UV irradiation was suppressed [25]. Similarly, in the ear tissue of mice coated with an emulsion containing an MAA (porphyra-334 or mycosporine-2-glycine), the decrease in the amounts of antioxidant enzymes during UV irradiation was suppressed, and the activities were maintained [26]. It has also been reported that shinorine and porphyra-334 had a promoting effect on the expression of antioxidant genes in human cells. These MAAs acted as antagonists of the binding between Kelch-like ECH-associated protein 1 (Keap1) and nuclear factor-erythroid 2-related factor 2 (Nrf2). As a result, they activated the Keap1 / Nrf2 signaling pathway, which is the main intracellular protective response pathway to oxidative stress [10].

Table 2. Effects of MAAs on the antioxidant system.

MAAs (Treatment Method)	Target Cells/Tissue	Effects of MAAs
Mycosporine-2-glycine (Added to the medium) [23]	RAW 264.7 macrophage cells	The addition of mycosporine-2-glycine suppressed the increase in the expression of *cat*, *sod1*, and *nrf1* genes caused by oxidative stress using hydrogen peroxide.
Mycosporine-2-glycine (Introduction of synthetic genes derived from the halotolerant cyanobacterium) [24]	The cyanobacterium *Synechococcus elongatus* PCC7942	The introduction of mycospoorine-2-glycine promoted the upregulation of *cat*, *sod1*, and *tpxA* genes by oxidative stress using hydrogen peroxide. (The *tpxA* gene encodes thioredoxin peroxidase, a homolog of PRDX).
Shinorine, Porphyra-334 (Application of an aqueous solution containing MAAs) [25]	Mouse back skin	Treatment with an aqueous solution containing MAAs suppressed the decrease in the activity of CAT, SOD, and GPX caused by UV irradiation stress. In addition, the production of malondialdehyde (MDA) was suppressed. MDA is one of the degradation products of lipid peroxidation and is a major marker of lipid peroxidation.
Porphyra-334, Mycosporine---glycine (Application of an emulsion containing each MAA) [26]	Mouse ear skin	The application of the emulsion suppressed the decrease in the accumulation and activities of CAT and SOD caused by UV irradiation stress.

(Table 2) cont.....

MAAs (Treatment Method)	Target Cells/Tissue	Effects of MAAs
Shinorine, Porphyra-334 (Added to the medium) [10]	Human skin fibroblasts	MAAs acted as antagonists of the binding between Keap1 and Nrf2, resulting in the activation of the Keap1/Nrf2 signaling pathway.

CONCLUDING REMARKS

This chapter summarized the antioxidative activities of MAAs. Various MAAs with non-enzymatic antioxidative activities have been reported. To date, although the molecular mechanisms for the antioxidative activities of MAAs are not well understood, they are thought to involve a conjugated system in the structures of MAAs [23]. In the future, further detailed analysis will be required to clarify the relationship between the molecular structure of MAAs and their antioxidative properties. Several effects of MAAs on the endogenous enzymatic antioxidant system have also been reported. These reports are very interesting from the perspective of the application of MAAs as cosmetics and pharmaceuticals.

LIST OF ABBREVIATIONS

ABTS 2 2'-azino-di-(3-ethylbenzothiazoline sulfonic acid) radical

CAT catalase

DPPH 1 1-diphenyl-2-picrylhydrazyl

FTC ferric thiocyanate

GPX glutathione peroxidase

GR glutathione reductase

MAA mycosporine-like amino acid

ORAC oxygen radical absorbance capacity

PC assay phosphatidylcholine peroxidation inhibition assay

PRDX peroxiredoxin

SOD superoxide dismutase

TBA thiobarbituric acid

TRXR thioredoxin reductase

ROS reactive oxygen species

UV ultraviolet

CONSENT FOR PUBLICATION

Not applicable.

CONFLICT OF INTEREST

The author declares no conflict of interest, financial or otherwise.

ACKNOWLEDGEMENTS

Declared none.

REFERENCES

[1] Sakurai, H.; Yasui, H.; Yamada, Y.; Nishimura, H.; Shigemoto, M. Detection of reactive oxygen species in the skin of live mice and rats exposed to UVA light: A research review on chemiluminescence and trials for UVA protection. *Photochem. Photobiol. Sci.,* **2005**, *4*(9), 715-720. [http://dx.doi.org/10.1039/b417319h] [PMID: 16121282]

[2] Masaki, H.; Okano, Y.; Sakurai, H. Generation of active oxygen species from advanced glycation end-products (AGEs) during ultraviolet light A (UVA) irradiation and a possible mechanism for cell damaging. *Biochim. Biophys. Acta, Gen. Subj.,* **1999**, *1428*(1), 45-56. [http://dx.doi.org/10.1016/S0304-4165(99)00056-2] [PMID: 10366759]

[3] Glady, A.; Tanaka, M.; Moniaga, C.S.; Yasui, M.; Hara-Chikuma, M. Involvement of NADPH oxidase 1 in UVB-induced cell signaling and cytotoxicity in human keratinocytes. *Biochem. Biophys. Rep.,* **2018**, *14*, 7-15. [http://dx.doi.org/10.1016/j.bbrep.2018.03.004] [PMID: 29872728]

[4] Ma, Q. Role of nrf2 in oxidative stress and toxicity. *Annu. Rev. Pharmacol. Toxicol.,* **2013**, *53*(1), 401-426. [http://dx.doi.org/10.1146/annurev-pharmtox-011112-140320] [PMID: 23294312]

[5] Shindo, Y.; Witt, E.; Han, D.; Epstein, W.; Packer, L. Enzymic and non-enzymic antioxidants in epidermis and dermis of human skin. *J. Invest. Dermatol.,* **1994**, *102*(1), 122-124. [http://dx.doi.org/10.1111/1523-1747.ep12371744] [PMID: 8288904]

[6] de la Coba, F.; Aguilera, J.; Figueroa, F.L.; de Gálvez, M.V.; Herrera, E. Antioxidant activity of mycosporine-like amino acids isolated from three red macroalgae and one marine lichen. *J. Appl. Phycol.,* **2009**, *21*(2), 161-169. [http://dx.doi.org/10.1007/s10811-008-9345-1]

[7] Ngoennet, S.; Nishikawa, Y.; Hibino, T.; Waditee-Sirisattha, R.; Kageyama, H. A method for the isolation and characterization of mycosporine-like amino acids from cyanobacteria. *Methods Protoc.,* **2018**, *1*(4), 46. [http://dx.doi.org/10.3390/mps1040046] [PMID: 31164584]

[8] Suh, S.S.; Hwang, J.; Park, M.; Seo, H.; Kim, H.S.; Lee, J.; Moh, S.; Lee, T.K. Anti-inflammation activities of mycosporine-like amino acids (MAAs) in response to UV radiation suggest potential anti-skin aging activity. *Mar. Drugs,* **2014**, *12*(10), 5174-5187. [http://dx.doi.org/10.3390/md12105174] [PMID: 25317535]

[9] Cheewinthamrongrod, V.; Kageyama, H.; Palaga, T.; Takabe, T.; Waditee-Sirisattha, R. DNA damage protecting and free radical scavenging properties of mycosporine-2-glycine from the Dead Sea cyanobacterium in A375 human melanoma cell lines. *J. Photochem. Photobiol. B,* **2016**, *164*, 289-295. [http://dx.doi.org/10.1016/j.jphotobiol.2016.09.037] [PMID: 27718421]

[10] Gacesa, R.; Lawrence, K.P.; Georgakopoulos, N.D.; Yabe, K.; Dunlap, W.C.; Barlow, D.J.; Wells, G.; Young, A.R.; Long, P.F. The mycosporine-like amino acids porphyra-334 and shinorine are antioxidants and direct antagonists of Keap1-Nrf2 binding. *Biochimie,* **2018**, *154*, 35-44. [http://dx.doi.org/10.1016/j.biochi.2018.07.020] [PMID: 30071261]

[11] Dunlap, W.C.; Yamamoto, Y. Small-molecule antioxidants in marine organisms: Antioxidant activity

of mycosporine-glycine. *Comp. Biochem. Physiol. B Biochem. Mol. Biol.,* **1995**, *112*(1), 105-114. [http://dx.doi.org/10.1016/0305-0491(95)00086-N]

[12] Suh, H.J.; Lee, H.W.; Jung, J. Mycosporine glycine protects biological systems against photodynamic damage by quenching singlet oxygen with a high efficiency. *Photochem. Photobiol.,* **2003**, *78*(2), 109-113. [http://dx.doi.org/10.1562/0031-8655(2003)078<0109:MGPBSA>2.0.CO;2] [PMID: 12945577]

[13] Saha, S.; Sen, A.; Mandal, S.; Adhikary, S.P.; Rath, J. Mycosporine-alanine, an oxo-mycosporine, protect *Hassallia byssoidea* from high UV and solar irradiation on the stone monument of Konark. *J. Photochem. Photobiol. B,* **2021**, *224*, 112302. [http://dx.doi.org/10.1016/j.jphotobiol.2021.112302] [PMID: 34537544]

[14] Nazifi, E.; Wada, N.; Asano, T.; Nishiuchi, T.; Iwamuro, Y.; Chinaka, S.; Matsugo, S.; Sakamoto, T. Characterization of the chemical diversity of glycosylated mycosporine-like amino acids in the terrestrial cyanobacterium *Nostoc commune. J. Photochem. Photobiol. B,* **2015**, *142*, 154-168. [http://dx.doi.org/10.1016/j.jphotobiol.2014.12.008] [PMID: 25543549]

[15] Whitehead, K.; Hedges, J.I. Photodegradation and photosensitization of mycosporine-like amino acids. *J. Photochem. Photobiol. B,* **2005**, *80*(2), 115-121. [http://dx.doi.org/10.1016/j.jphotobiol.2005.03.008] [PMID: 15893470]

[16] Nishida, Y.; Kumagai, Y.; Michiba, S.; Yasui, H.; Kishimura, H. Efficient extraction and antioxidant capacity of mycosporine-like amino acids from red alga dulse *Palmaria palmata* in Japan. *Mar. Drugs,* **2020**, *18*(10), 502. [http://dx.doi.org/10.3390/md18100502] [PMID: 33008002]

[17] Lawrence, K.P.; Gacesa, R.; Long, P.F.; Young, A.R. Molecular photoprotection of human keratinocytes *in vitro* by the naturally occurring mycosporine-like amino acid palythine. *Br. J. Dermatol.,* **2018**, *178*(6), 1353-1363. [http://dx.doi.org/10.1111/bjd.16125] [PMID: 29131317]

[18] Nakayama, R.; Tamura, Y.; Kikuzaki, H.; Nakatani, N. Antioxidant effect of the constituents of susabinori *(Porphyra yezoensis). J. Am. Oil Chem. Soc.,* **1999**, *76*(5), 649-653. [http://dx.doi.org/10.1007/s11746-999-0017-3]

[19] Moliné, M.; Arbeloa, E.M.; Flores, M.R.; Libkind, D.; Farías, M.E.; Bertolotti, S.G.; Churio, M.S.; van Broock, M.R. UVB photoprotective role of mycosporines in yeast: photostability and antioxidant activity of mycosporine-glutaminol-glucoside. *Radiat. Res.,* **2011**, *175*(1), 44-50. [http://dx.doi.org/10.1667/RR2245.1] [PMID: 21175346]

[20] Matsui, K.; Nazifi, E.; Kunita, S.; Wada, N.; Matsugo, S.; Sakamoto, T. Novel glycosylated mycosporine-like amino acids with radical scavenging activity from the cyanobacterium *Nostoc commune. J. Photochem. Photobiol. B,* **2011**, *105*(1), 81-89. [http://dx.doi.org/10.1016/j.jphotobiol.2011.07.003] [PMID: 21813286]

[21] Sakamoto, T.; Hashimoto, A.; Yamaba, M.; Wada, N.; Yoshida, T.; Inoue-Sakamoto, K.; Nishiuchi, T.; Matsugo, S. Four chemotypes of the terrestrial cyanobacterium *Nostoc commune* characterized by differences in the mycosporine□like amino acids. *Phycol. Res.,* **2019**, *67*(1), 3-11. [http://dx.doi.org/10.1111/pre.12333]

[22] Ishihara, K.; Watanabe, R.; Uchida, H.; Suzuki, T.; Yamashita, M.; Takenaka, H.; Nazifi, E.; Matsugo, S.; Yamaba, M.; Sakamoto, T. Novel glycosylated mycosporine-like amino acid, 13- O -(- -galactosyl)-porphyra-334, from the edible cyanobacterium Nostoc sphaericum -protective activity on human keratinocytes from UV light. *J. Photochem. Photobiol. B,* **2017**, *172*, 102-108. [http://dx.doi.org/10.1016/j.jphotobiol.2017.05.019] [PMID: 28544967]

[23] Tarasuntisuk, S.; Palaga, T.; Kageyama, H.; Waditee-Sirisattha, R. Mycosporine-2-glycine exerts anti-inflammatory and antioxidant effects in lipopolysaccharide (LPS)-stimulated RAW 264.7 macrophages. *Arch. Biochem. Biophys.,* **2019**, *662*, 33-39. [http://dx.doi.org/10.1016/j.abb.2018.11.026] [PMID: 30502329]

[24] Pingkhanont, P.; Tarasuntisuk, S.; Hibino, T.; Kageyama, H.; Waditee-Sirisattha, R. Expression of a stress-responsive gene cluster for mycosporine-2-glycine confers oxidative stress tolerance in *Synechococcus elongatus* PCC7942. *FEMS Microbiol. Lett.,* **2019**, *366*(9), fnz115. [http://dx.doi.org/10.1093/femsle/fnz115] [PMID: 31132117]

[25] Ying, R.; Zhang, Z.; Zhu, H.; Li, B.; Hou, H. The Protective Effect of Mycosporine-Like Amino Acids (MAAs) from *Porphyra yezoensis* in a Mouse Model of UV Irradiation-Induced Photoaging. *Mar. Drugs,* **2019**, *17*(8), 470. [http://dx.doi.org/10.3390/md17080470] [PMID: 31416181]

[26] Waditee-Sirisattha, R.; Kageyama, H. Protective effects of mycosporine-like amino acid-containing emulsions on UV-treated mouse ear tissue from the viewpoints of antioxidation and antiglycation. *J. Photochem. Photobiol. B,* **2021**, *223*, 112296. [http://dx.doi.org/10.1016/j.jphotobiol.2021.112296] [PMID: 34450363]

CHAPTER 7

Biological Activities of MAAs and their Applications 3: Anti-inflammatory Effects

Abstract: Inflammation is the defensive reaction system that occurs when the body receives a harmful stimulus and tries to remove it. In general, the area where the reaction occurs has a fever, swelling, redness, and pain. The stimuli that cause inflammation are diverse but include UV irradiation and reactive oxygen species. This chapter briefly describes the inflammatory response pathways caused by these stimuli. After that, it outlines the effects of mycosporine-like amino acids (MAAs) on the accumulation, activity, and regulation of factors contained in the inflammatory pathway. Although research findings are accumulating, the molecular mechanisms are still unknown. Details of the relationship between the molecular structures of MAAs and their functions in the inflammatory pathway await further study.

Keywords: Antioxidant, Cyclooxygenase-2, Interleukin-1, Interleukin-6, Inducible NO synthase, Inflammation, Inhibitor protein of NF-κB, Lipopolysaccharides, Mycosporine-like amino acid, Nuclear factor-kappa B, Nitrogen monoxide, Prostaglandin E_2, Reactive oxygen species, Tumor necrosis factor α, Ultraviolet.

INTRODUCTION

Inflammation is a physiological defense mechanism against molecular and cellular damage caused by various stresses, including irradiation, oxidative stress, infection, and exposure to endotoxins, such as lipopolysaccharides (LPS) [1]. UV irradiation is known to induce inflammation. Distinct patterns of inflammation caused by UV irradiation are caused by exposure to rays of a specific wavelength. The three groups, UV-A, UV-B, and UV-C, are categorized based on these different inflammatory patterns [2]. In particular, the erythema caused by UV-B irradiation is called sunburn. It is well known that the inflammatory response induced by UV-B irradiation is associated with a variety of factors. These factors include nitrogen monoxide (NO), inducible NO synthase (iNOS), prostaglandin E2 (PGE2), cyclooxygenase-2 (COX-2), tumor necrosis factor (TNF-α), and cytokines such as interleukin-1 (IL-1) and interleukin-6 (IL-6) (Fig. **1**). These molecules are produced mainly in keratinocytes, the major cell type of the epider-

Hakuto Kageyama

mis, and are regulated by nuclear factor-kappa B (NF-κB) [3]. NF-κB regulates the expression of genes involved in inflammation, oxidative stress response, differentiation, proliferation, apoptosis, and cell adhesion [4, 5]. In the cytoplasm, the NF-κB protein interacts with the inhibitor protein of NF-κB (IκB) and is inactivated. When IκB is degraded in response to stress stimuli, activated NF-κB translocates to the nucleus [4, 6]. It is known that iNOS and COX-2 are involved in the molecular response to inflammatory stimuli. *iNOS* gene expression is induced as inflammation progresses, followed by overproduction of the pro-inflammatory factor NO. It has also been reported that the expression of COX-2, which is involved in the production of PGE2, was induced by UV-B irradiation in the human skin tissue and cultured human keratinocytes [7]. PGE2 is a bioactive lipid associated with the induction of inflammation and cancer. In addition, reactive oxygen species (ROS) are associated with the inflammatory response. In fact, it has been found that COX-2 expression is induced by ROS in various cells [8].

Anti-inflammatory molecules have been widely studied not only for skin care applications, but also for the treatment of chronic inflammatory diseases including rheumatoid arthritis, psoriasis, chronic obstructive pulmonary disease, multiple sclerosis, and inflammatory bowel disease. Anti-inflammatory compounds are also useful in the treatment of cardiovascular diseases such as atherosclerosis and neurodegenerative diseases such as Parkinson's disease.

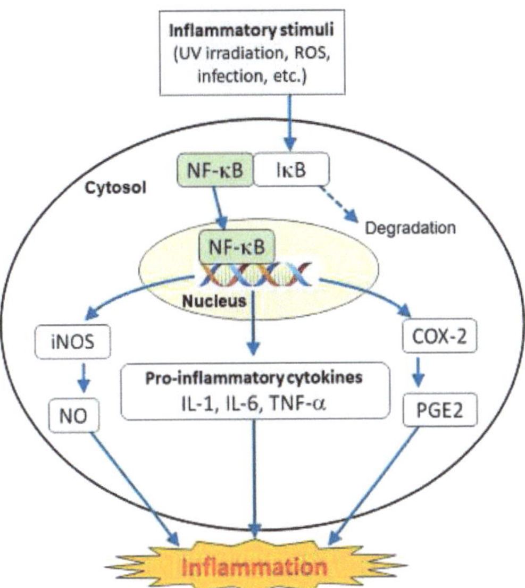

Fig. (1). Inflammatory response induced by the stimuli, including UV-irradiation. The NF-κB protein activated *via* inflammatory stimuli can translocate into the nucleus and regulate genes involved in the inflammatory response. See the text for further explanation.

EFFECTS OF MAAS ON UV-B-INDUCED INFLAMMATORY PATHWAY

So far, the results of studies on the anti-inflammatory effect of mycosporine-like amino acids (MAAs), including shinorine, porphyra-334, mycosporine-glycine, and mycosporine-2-glycine, have been reported (Table **1**). Suh *et al.* showed that the upregulation of *COX-2* gene expression in immortalized human keratinocyte HaCaT cells by UV irradiation was suppressed by shinorine or mycosporine-glycine but not porphyra-334 [9]. However, Becker *et al.* reported that porphyra-334 suppressed LPS-induced NF-κB activity in human myelomonocyte THP--‑Blue cells. In contrast, shinorine promoted LPS-induced NF-κB activity [10]. Ying *et al.* showed that the expression levels of TNF-α, IL-1β, and IL-6 were reduced in the skin tissue of mice treated with a mixture of shinorine and porphyra-334. Treatment of MAAs at a concentration of 20 mg/ml reduced mRNA levels of NF-κB, IL-1β, and IL-6 by 21.05%, 25.10%, and 42.13%, respectively [11]. In addition, Tarasuntisuk *et al.* revealed that mycosporine--‑glycine significantly inhibited NO production in LPS-stimulated RAW264.7 macrophage cells. Although mycosporine-glycine, palythine, shinorine, and porphyra-334 showed a maximum inhibitory effect of about 20%, mycosporine--‑glycine showed an inhibitory effect of about 50% at 10 μM. The relationship between the molecular structures of MAAs and their inhibitory effects is still unclear, but it is an interesting research subject. Besides, it was reported that mycosporine-2-glycine strongly suppressed the expression of *iNOS* and *COX-2* genes [12]. Cheewinthamrongrod *et al.* reported that oxidative stress-induced NF-κB accumulation in human skin fibroblasts was suppressed by the addition of mycosporine-2-glycine [13]. Given that antioxidant molecules such as vitamin E is shown to inhibit the NF-κB signaling pathway [14], it is possible that the strong antioxidative capacity of mycosporine-2-glycine is related to this phenomenon.

Table 1. Effects of MAAs on UV-B-induced inflammatory pathways.

MAAs (Treatment)	Target Cells or Tissue	The Effects of MAAs
Shinorine, Mycosporine-glycine (Addition to the medium) [9]	HaCaT cells	The upregulation of *COX-2* gene expression caused by UV irradiation stress was suppressed.
Porphyra-334 (Addition to the medium) [10]	THP-1-Blue cells	The induction of NF-κB activity caused by LPS was suppressed.
Shinorine (Addition to the medium) [10]	THP-1-Blue cells	The induction of NF-κB activity caused by LPS was promoted.
Mixture of shinorine and porphyra-334 (Application of an aqueous solution containing MAAs) [11]	Mouse back skin	The induction of NF-κB activity caused by UV irradiation was suppressed.

(Table 1) cont.....

MAAs (Treatment)	Target Cells or Tissue	The Effects of MAAs
Mycosporine-2-glycine (Addition to the medium) [12]	RAW 264.7 macrophage cells	The induction of NO production caused by LPS was suppressed. The expression of *iNOS* and *COX-2* genes was significantly suppressed.
Mycosporine-2-glycine (Addition to the medium) [13]	Human skin fibroblast cells	The induction of NF-κB accumulation caused by oxidative stress using hydrogen peroxide was suppressed.

CONCLUDING REMARKS

This chapter described the effects of MAAs exerted on the inflammatory pathway. Some MAAs appear to have anti-inflammatory effects. Various effects have been shown depending on the types of MAAs used in the investigations. However, the relationship between the molecular structure of each MAA and each effect is still unclear. Because the types of MAAs studied so far have been limited, research should be conducted using more diverse MAAs in the future.

LIST OF ABBREVIATIONS

COX-2 cyclooxygenase-2

IL interleukin

iNOS inducible NO synthase

IκB inhibitor protein of NF-κB

LPS lipopolysaccharides

MAA mycosporine-like amino acid

NF-κB nuclear factor-kappa B

NO nitrogen monoxide

PGE2 prostaglandin E_2

ROS reactive oxygen species

TNF-α tumor necrosis factor α

UV ultraviolet

CONSENT FOR PUBLICATION

Not applicable.

CONFLICT OF INTEREST

The author declares no conflict of interest, financial or otherwise.

ACKNOWLEDGEMENTS

Declared none.

REFERENCES

[1] Newton, K.; Dixit, V.M. Signaling in innate immunity and inflammation. *Cold Spring Harb. Perspect. Biol.,* **2012**, *4*(3), a006049.
[http://dx.doi.org/10.1101/cshperspect.a006049] [PMID: 22296764]

[2] Hruza, L.L.; Pentland, A.P. Mechanisms of UV-induced inflammation. *J. Invest. Dermatol.,* **1993**, *100*(1), S35-S41.
[http://dx.doi.org/10.1038/jid.1993.21] [PMID: 8423392]

[3] Prasad, N.; Radhiga, T.; Agilan, B.; Muzaffer, U.; Karthikeyan, R.; Kanimozhi, G.; Paul, V.I. Phytochemicals as modulators of ultraviolet-b radiation induced cellular and molecular events: A review. *Journal of Radiation and Cancer Research,* **2016**, *7*(1), 2.
[http://dx.doi.org/10.4103/0973-0168.184607]

[4] Gupta, S.C.; Sundaram, C.; Reuter, S.; Aggarwal, B.B. Inhibiting NF-κB activation by small molecules as a therapeutic strategy. *Biochim. Biophys. Acta. Gene Regul. Mech.,* **2010**, *1799*(10-12), 775-787.
[http://dx.doi.org/10.1016/j.bbagrm.2010.05.004] [PMID: 20493977]

[5] Ambrozova, N.; Ulrichova, J.; Galandakova, A. Models for the study of skin wound healing. The role of Nrf2 and NF-κB. *Biomed. Pap. Med. Fac. Univ. Palacky Olomouc Czech Repub.,* **2017**, *161*(1), 1-13.
[http://dx.doi.org/10.5507/bp.2016.063] [PMID: 28115750]

[6] Ghosh, S.; May, M.J.; Kopp, E.B. NF-kappa B and Rel proteins: evolutionarily conserved mediators of immune responses. *Annu. Rev. Immunol.,* **1998**, *16*(1), 225-260.
[http://dx.doi.org/10.1146/annurev.immunol.16.1.225] [PMID: 9597130]

[7] Bowden, G.T. Prevention of non-melanoma skin cancer by targeting ultraviolet-B-light signalling. *Nat. Rev. Cancer,* **2004**, *4*(1), 23-35.
[http://dx.doi.org/10.1038/nrc1253] [PMID: 14681688]

[8] Onodera, Y.; Teramura, T.; Takehara, T.; Shigi, K.; Fukuda, K. Reactive oxygen species induce Cox-2 expression *via* TAK1 activation in synovial fibroblast cells. *FEBS Open Bio,* **2015**, *5*(1), 492-501.
[http://dx.doi.org/10.1016/j.fob.2015.06.001] [PMID: 26110105]

[9] Suh, S.S.; Hwang, J.; Park, M.; Seo, H.; Kim, H.S.; Lee, J.; Moh, S.; Lee, T.K. Anti-inflammation activities of mycosporine-like amino acids (MAAs) in response to UV radiation suggest potential anti-skin aging activity. *Mar. Drugs,* **2014**, *12*(10), 5174-5187.
[http://dx.doi.org/10.3390/md12105174] [PMID: 25317535]

[10] Becker, K.; Hartmann, A.; Ganzera, M.; Fuchs, D.; Gostner, J. Immunomodulatory effects of the mycosporine-like amino acids shinorine and porphyra-334. *Mar. Drugs,* **2016**, *14*(6), 119.
[http://dx.doi.org/10.3390/md14060119] [PMID: 27338421]

[11] Ying, R.; Zhang, Z.; Zhu, H.; Li, B.; Hou, H. The protective effect of mycosporine-like amino acids (MAAs) from *Porphyra yezoensis* in a mouse model of UV irradiation-induced photoaging. *Mar. Drugs,* **2019**, *17*(8), 470.
[http://dx.doi.org/10.3390/md17080470] [PMID: 31416181]

[12] Tarasuntisuk, S.; Palaga, T.; Kageyama, H.; Waditee-Sirisattha, R. Mycosporine-2-glycine exerts anti-inflammatory and antioxidant effects in lipopolysaccharide (LPS)-stimulated RAW 264.7 macrophages. *Arch. Biochem. Biophys.,* **2019**, *662*, 33-39.
[http://dx.doi.org/10.1016/j.abb.2018.11.026] [PMID: 30502329]

[13] Cheewinthamrongrod, V.; Kageyama, H.; Palaga, T.; Takabe, T.; Waditee-Sirisattha, R. DNA damage

protecting and free radical scavenging properties of mycosporine-2-glycine from the Dead Sea cyanobacterium in A375 human melanoma cell lines. *J. Photochem. Photobiol. B,* **2016**, *164*, 289-295. [http://dx.doi.org/10.1016/j.jphotobiol.2016.09.037] [PMID: 27718421]

[14] Ginn-Pease, M.E.; Whisler, R.L. Redox signals and NF-kappaB activation in T cells. *Free Radic. Biol. Med.,* **1998**, *25*(3), 346-361. [http://dx.doi.org/10.1016/S0891-5849(98)00067-7] [PMID: 9680181]

CHAPTER 8

Biological Activities of MAAs and their Applications 4: Anti-glycative Properties

Abstract: Advanced glycation end products (AGEs) are formed by a series of chemical reactions initiated by non-enzymatic glycation reactions. In this process, the reducing sugar binds to the free amino group of the protein. The formation of AGEs that accompany the aging process is thought to be associated with various diseases such as diabetes and Alzheimer's disease. A number of inhibitors derived from synthetic compounds and natural products have been developed and evaluated to prevent the formation of AGEs. Compared to synthetic compounds, natural products are considered to be relatively safe for human consumption, so there is an increasing demand for compounds derived from natural products. From this perspective, this chapter focuses on mycosporine-like amino acids as naturally occurring inhibitors against AGEs formation.

Keywords: Advanced glycation end products, Amino guanidine, Glycation, Maillard reaction, Natural inhibitors, Reactive oxygen species, Synthetic inhibitors.

INTRODUCTION

Glycation is a series of chemical reactions involving the non-enzymatic binding of sugar molecules to free amino groups and hydroxy groups of biocompounds such as proteins and lipids, which is also called the Maillard reaction [1]. The Maillard reaction was discovered in 1912 by the French chemist Louis-Camille Maillard as a reaction associated with browning during the cooking and storage of food [2]. The final product of the glycation reaction between the free amino group of a protein and a reducing sugar such as glucose is called an advanced glycation end product (AGE) [3]. Accumulation of AGEs impairs the structure and function of tissue proteins in the body and is a complication of blood vessels and kidneys in diabetic patients, atherosclerosis, and Alzheimer's disease [1, 4 - 6]. For example, high levels of AGEs have been detected in the blood and tissues of diabetics, which leads to complications such as nephropathy and neuropathy [7, 8]. It is thought that a decrease in enzyme activity due to changes in the charge state and the formation of crosslinks by the glycation of proteins is related to the adverse

Hakuto Kageyama

effects of these AGEs. Changes in protein properties can affect a variety of biological phenomena, including tissue homeostasis and regulation of gene expression. In addition, because there are many structures of AGEs, it is highly possible that many proteins are modified even if the amount of each AGE structure accumulated is small.

AGEs are also associated with the aging process, and the amount of AGEs accumulated in the body increases with normal aging. Thus, endogenously formed AGEs are a common product of metabolism, but it is also known that AGE formation is enhanced by oxidative stress [9]. A number of studies have shown that elevated levels of reactive oxygen species (ROS) caused by oxidative stress are involved in the development of diabetes and its complications [10].

AGEs are generated through a complicated multi-step reaction that can be roughly divided into two stages (Fig. **1**). In the reaction process of the early stage, the electrophilic carbonyl group (R_1–C(=O)–R_2) in a reduced sugar reacts with the free amino group (–NH_2) in the N-terminal of the protein and the side chains of the basic amino acid residues lysine, arginine, and histidine [11]. Subsequently, an unstable Schiff base (R_1R_2C=N–R_3) is formed [12]. The rearrangement of this compound results in the formation of a stable Amadori product. Then, in the late reaction process, the Amadori product irreversibly binds to an amino acid residue of the protein to form a cross-linked product [12, 13]. Further oxidation, dehydration, polymerization, and oxidative decomposition of Amadori products produce a variety of AGEs [14].

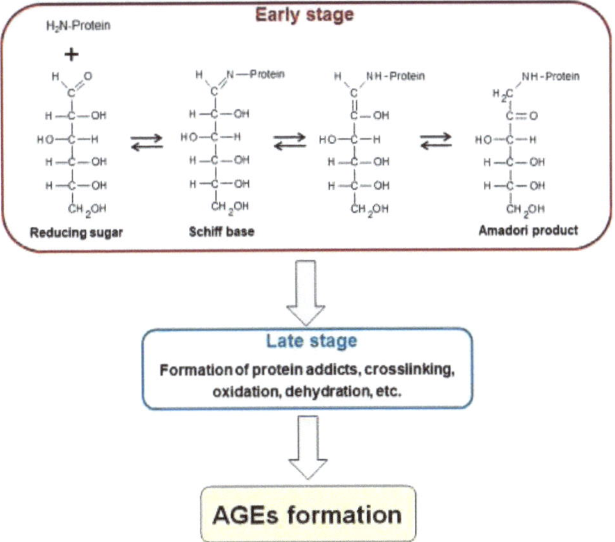

Fig. (1). The formation of AGEs by the glycation reaction.

At present, more than 40 types of structures of AGEs have been reported [15], [16]. Some of them are crosslinked structures, such as pentosidine, glucosepane, crossline, glyoxal-lysine dimer (GOLD), and methylglyoxal-lysine dimer (MOLD). Pentosidine and glucosepane are AGEs that crosslink lysine and arginine. In contrast, crossline, GOLD, and MOLD crosslink two lysine residues. Others are non-crosslinked AGE adducts to the protein, such as N^{ε}-(carboxymethyl)lysine (CML), N^{ε}-(carboxyethyl)lysine (CEL), and pyrraline (Fig. **2**).

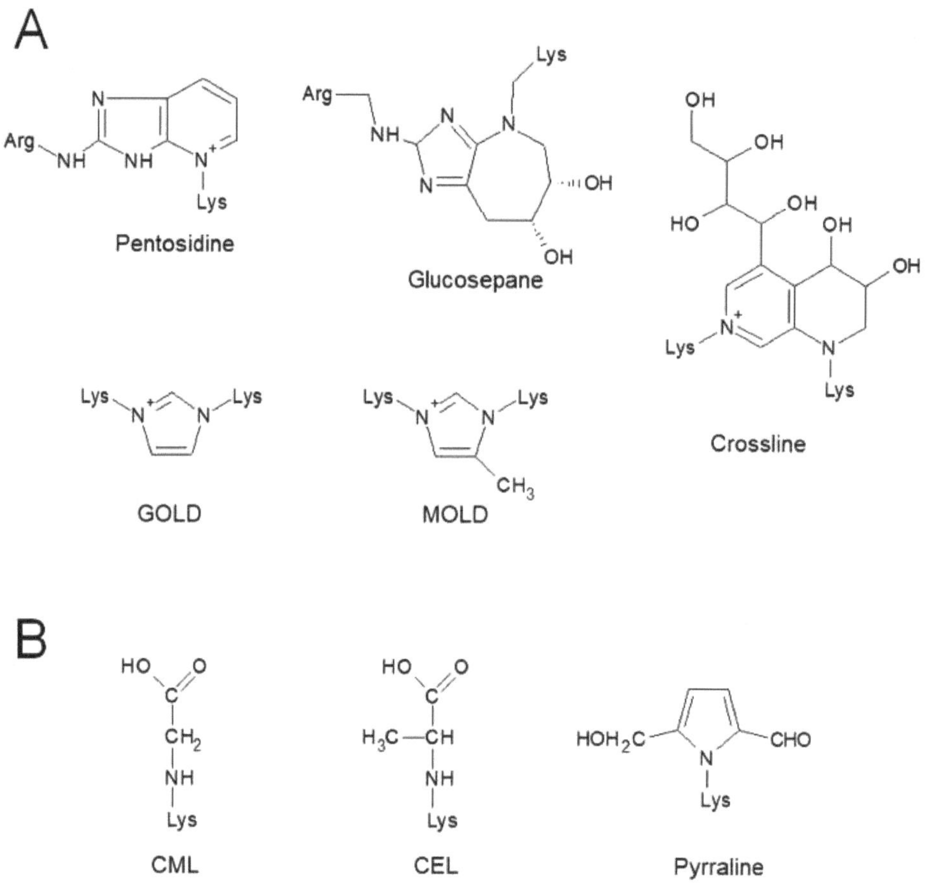

Fig. (2). Schematic representation of structures of cross-linked type AGEs (A) and non-crosslinked AGE adducts (B).

Inhibitors of various glycation reactions have been derived from synthetic compounds to suppress the formation of AGEs. An example of a well-known synthetic inhibitor is aminoguanidine, which has a nucleophilic hydrazine [17]

(Fig. **3**). Aminoguanidine was found to have an inhibitory effect on diabetes-related AGE formation. Besides aminoguanidine, pyridoxamine, a derivative of vitamin B6, is also known as an AGE inhibitor [18] (Fig. **3**). Pyridoxamine and aminoguanidine suppress AGE formation by reacting with Amadori products. Although the detailed mechanism has not been clarified, it has been suggested that the metal chelating activities of these compounds have the potential to inhibit AGE formation. This is because all analogs of these compounds with AGE inhibitory activity also had metal chelating activity [19]. However, the administration of aminoguanidine to patients induced pernicious anemia and the development of antinuclear antibodies [20, 21]. Animal studies also showed that the administration of high concentrations of aminoguanidine induced pancreatic and renal tumors [22]. Eventually, the phase III clinical trial of aminoguanidine as a treatment for patients with type I diabetes was discontinued. Thus, synthetic compounds can cause serious side effects. However, it was reported that a phase II clinical trial assessing the therapeutic effects of pyridoxamine in diabetic neuropathy yielded promising data [23]. In addition to aminoguanidine and pyridoxamine, a number of other synthetic compounds, including diclofenac and acetylsalicylic acid (Fig. **3**), have been developed.

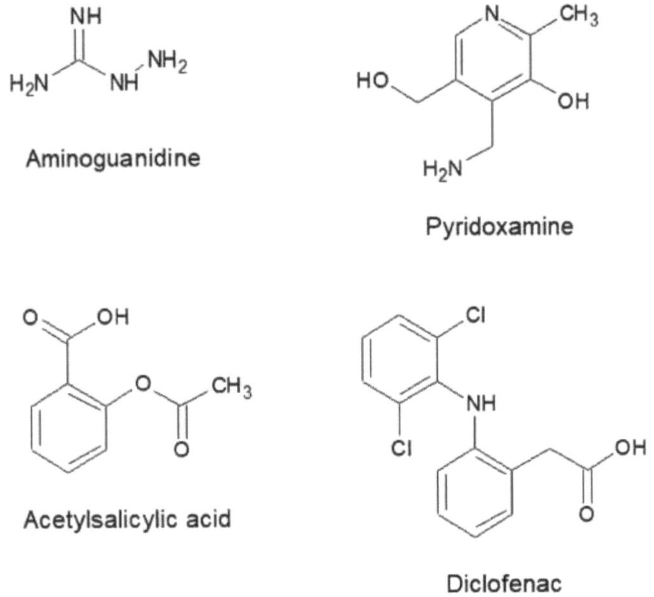

Aminoguanidine

Pyridoxamine

Acetylsalicylic acid

Diclofenac

Fig. (3). Chemical structures of representative synthetic inhibitors against glycation.

As mentioned above, synthetic compounds can have serious side effects, and the same is true for substances derived from natural products. However, due to the high diversity and functionality of natural products, safe and highly active inhibitors of saccharification may be found. In particular, several MAAs have been reported as promising natural inhibitors of glycation [24, 25].

ANTI-GLYCATIVE PROPERTIES OF MAAS

So far, 11 types of MAAs have been reported to have anti-glycative activities *in vitro* (Table **1**). Mycosporine-2-glycine purified from the halotolerant cyanobacterium *Halothece* sp. PCC7418 inhibited the cross-linking reaction in the glycation of hen egg white lysozyme [26]. The half-maximal (50%) inhibitory concentration (IC_{50}) value was 1.6 mM, which was smaller than the IC_{50} value of 4.7 mM for aminoguanidine. In addition, shinorine, porphyra-334, mycosporine-methylamin--threonine, mycosporine-alanine-glycine, aplysiapalythine A, asterina-330, palythine, bostrychine D, bostrychine E, and bostrychine F were reported to inhibit the glycation reaction of bovine serum albumin (BSA) [27]. In particular, shinorine, porphyra-334, mycosporine-alanine-glycine, and bostrychine D showed a high inhibitory activity. There has also been a report that porphyra-334 purified from Helioguard 365 inhibited the glycation reaction of elastin [28]. In that study, the IC_{50} value of porphyra-334 was 7.6 mM, which was equivalent to the IC_{50} value of aminoguanidine, 7.4 mM.

Table 1. MAAs with anti-glycative activities.

MAAs	Proteins and Reducing Sugars used in the Experiment	Inhibitory Activity (IC_{50})
Mycosporine-2-glycine [26]	Lysozyme + Ribose	1.6 mM
Shinorine [27]	BSA + Ribose	103 μM
Porphyra-334 [27]	BSA + Ribose	90 μM
Mycosporine-methylamine-threonine [27]	BSA + Ribose	150 μM
Mycosporine-alanine-glycine [27]	BSA + Ribose	75 μM
Aplysiapalythine A [27]	BSA + Ribose	400 μM
Asterina-330 [27]	BSA + Ribose	125 μM
Palythine [27]	BSA + Ribose	700 μM
Bostrychine D [27]	BSA + Ribose	85 μM
Bostrychine E [27]	BSA + Ribose	150 μM
Bostrychine F [27]	BSA + Ribose	200 μM
Porphyra-334 [28]	Elastin + Glyceraldehyde	7.6 mM

Although the molecular mechanism of the glycation-inhibitory activity of these MAAs is currently unknown, the antioxidative properties of MAAs might be correlated to glycation because oxidative stress is known to enhance the formation of AGEs. In addition, glycation is considered to be a source of ROS generated by oxidative and non-oxidative pathways [21]. Thus, antioxidant molecules, including MAAs, could exert an inhibitory effect against AGEs formation. The antioxidative properties of MAAs were summarized in Chapter 6. As described above, it has been reported that strong inhibitors of the glycation reaction such as aminoguanidine exhibit metal chelating activity. It has been suggested that MAAs also have a metal chelating ability, and this may be related to their anti-glycation ability. The chelating activity of MAAs is summarized in Chapter 10.

CONCLUDING REMARKS

The formation of AGEs is involved in various diseases and aging processes. In this chapter, after outlining the basics of the formation mechanism of AGEs, the anti-glycative activities of MAAs were described. Although the inhibitory mechanism has not yet been clarified, it will be of interest to industries including the pharmaceutical field as another useful biological activity of MAAs.

LIST OF ABBREVIATIONS

AGEs advanced glycation end products

CEL N^{ε}-(carboxyethyl)lysine

CML N^{ε}-(carboxymethyl)lysine

GOLD glyoxal-lysine dimer

MAA mycosporine-like amino acid

MOLD methylglyoxal-lysine dimer

ROS reactive oxygen species

CONSENT FOR PUBLICATION

Not applicable.

CONFLICT OF INTEREST

The author declares no conflict of interest, financial or otherwise.

ACKNOWLEDGEMENTS

Declared none.

REFERENCES

[1] Fournet, M.; Bonté, F.; Desmoulière, A. Glycation damage: A possible hub for major pathophysiological disorders and aging. *Aging Dis.,* **2018**, *9*(5), 880-900. [http://dx.doi.org/10.14336/AD.2017.1121] [PMID: 30271665]

[2] Maillard, L.C. Action des acides amines sur les sucres: formation des melanoidines par voie methodique. *Comptes Rendus de l'Academie des Science.,* **1912**, *154*, 66-68.

[3] Freund, M.A.; Chen, B.; Decker, E.A. The inhibition of advanced glycation end products by carnosine and other natural dipeptides to reduce diabetic and age-related complications. *Compr. Rev. Food Sci. Food Saf.,* **2018**, *17*(5), 1367-1378.
[http://dx.doi.org/10.1111/1541-4337.12376] [PMID: 33350165]

[4] Uribarri, J.; Cai, W.; Peppa, M.; Goodman, S.; Ferrucci, L.; Striker, G.; Vlassara, H. Circulating glycotoxins and dietary advanced glycation endproducts: two links to inflammatory response, oxidative stress, and aging. *J. Gerontol. A Biol. Sci. Med. Sci.,* **2007**, *62*(4), 427-433.
[http://dx.doi.org/10.1093/gerona/62.4.427] [PMID: 17452738]

[5] Matsuura, N.; Aradate, T.; Kurosaka, C.; Ubukata, M.; Kittaka, S.; Nakaminami, Y.; Gamo, K.; Kojima, H.; Ohara, M. Potent protein glycation inhibition of plantagoside in Plantago major seeds. *BioMed Res. Int.,* **2014**, *2014*, 1-5.
[http://dx.doi.org/10.1155/2014/208539] [PMID: 24895551]

[6] Münch, G.; Thome, J.; Foley, P.; Schinzel, R.; Riederer, P. Advanced glycation endproducts in ageing and Alzheimer's disease. *Brain Res. Brain Res. Rev.,* **1997**, *23*(1-2), 134-143.
[http://dx.doi.org/10.1016/S0165-0173(96)00016-1] [PMID: 9063589]

[7] Tanji, N.; Markowitz, G.S.; Fu, C.; Kislinger, T.; Taguchi, A.; Pischetsrieder, M.; Stern, D.; Schmidt, A.M.; D'Agati, V.D. Expression of advanced glycation end products and their cellular receptor RAGE in diabetic nephropathy and nondiabetic renal disease. *J. Am. Soc. Nephrol.,* **2000**, *11*(9), 1656-1666.
[http://dx.doi.org/10.1681/ASN.V1191656] [PMID: 10966490]

[8] Stracke, H.; Hammes, H.; Werkmann, D.; Mavrakis, K.; Bitsch, I.; Netzel, M.; Geyer, J.; Köpcke, W.; Sauerland, C.; Bretzel, R.; Federlin, K. Efficacy of benfotiamine *versus* thiamine on function and glycation products of peripheral nerves in diabetic rats. *Exp. Clin. Endocrinol. Diabetes,* **2001**, *109*(6), 330-336.
[http://dx.doi.org/10.1055/s-2001-17399] [PMID: 11571671]

[9] Mustafa, I.; Ahmad, S.; Dixit, K.; Moinuddin, ; Ahmad, J.; Ali, A. Glycated human DNA is a preferred antigen for anti-DNA antibodies in diabetic patients. *Diabetes Res. Clin. Pract.,* **2012**, *95*(1), 98-104.
[http://dx.doi.org/10.1016/j.diabres.2011.09.018] [PMID: 22001283]

[10] Gayathri, K.R.; Anitha, R.; Lakshmi, T. Inhibition of advanced glycation end-product formation by lutein from *Tagetes erecta. Pharmacogn. J.,* **2018**, *10*(4), 734-737.
[http://dx.doi.org/10.5530/pj.2018.4.123]

[11] Hipkiss, A.R.; Michaelis, J.; Syrris, P. Non-enzymatic glycosylation of the dipeptide L -carnosine, a potential anti-protein-cross-linking agent. *FEBS Lett.,* **1995**, *371*(1), 81-85.
[http://dx.doi.org/10.1016/0014-5793(95)00849-5] [PMID: 7664889]

[12] Paul, R.G.; Bailey, A.J. Glycation of collagen: the basis of its central role in the late complications of ageing and diabetes. *Int. J. Biochem. Cell Biol.,* **1996**, *28*(12), 1297-1310.
[http://dx.doi.org/10.1016/S1357-2725(96)00079-9] [PMID: 9022289]

[13] Ahmed, N. Advanced glycation endproducts—role in pathology of diabetic complications. *Diabetes Res. Clin. Pract.,* **2005**, *67*(1), 3-21.
[http://dx.doi.org/10.1016/j.diabres.2004.09.004] [PMID: 15620429]

[14] Thorpe, S.R.; Baynes, J.W. Maillard reaction products in tissue proteins: New products and new perspectives. *Amino Acids,* **2003**, *25*(3-4), 275-281.
[http://dx.doi.org/10.1007/s00726-003-0017-9] [PMID: 14661090]

[15] Xanthis, A.; Hatzitolios, A.; Koliakos, G.; Tatola, V. Advanced glycosylation end products and nutrition--a possible relation with diabetic atherosclerosis and how to prevent it. *J. Food Sci.,* **2007**, *72*(8), R125-R129.

[http://dx.doi.org/10.1111/j.1750-3841.2007.00508.x] [PMID: 17995617]

[16] Bilova, T.; Paudel, G.; Shilyaev, N.; Schmidt, R.; Brauch, D.; Tarakhovskaya, E.; Milrud, S.; Smolikova, G.; Tissier, A.; Vogt, T.; Sinz, A.; Brandt, W.; Birkemeyer, C.; Wessjohann, L.A.; Frolov, A. Global proteomic analysis of advanced glycation end products in the *Arabidopsis* proteome provides evidence for age-related glycation hot spots. *J. Biol. Chem.*, **2017**, *292*(38), 15758-15776. [http://dx.doi.org/10.1074/jbc.M117.794537] [PMID: 28611063]

[17] Brownlee, M.; Vlassara, H.; Kooney, A.; Ulrich, P.; Cerami, A. Aminoguanidine prevents diabetes-induced arterial wall protein cross-linking. *Science,* **1986**, *232*(4758), 1629-1632. [http://dx.doi.org/10.1126/science.3487117] [PMID: 3487117]

[18] Khalifah, R.G.; Chen, Y.; Wassenberg, J.J. Post-Amadori AGE inhibition as a therapeutic target for diabetic complications: a rational approach to second-generation Amadorin design. *Ann. N. Y. Acad. Sci.*, **2005**, *1043*(1), 793-806. [http://dx.doi.org/10.1196/annals.1333.092] [PMID: 16037307]

[19] Nagai, R.; Murray, D.B.; Metz, T.O.; Baynes, J.W. Chelation: a fundamental mechanism of action of AGE inhibitors, AGE breakers, and other inhibitors of diabetes complications. *Diabetes,* **2012**, *61*(3), 549-559. [http://dx.doi.org/10.2337/db11-1120] [PMID: 22354928]

[20] Nilsson, B.O. Biological effects of aminoguanidine: An update. *Inflamm. Res.*, **1999**, *48*(10), 509-515. [http://dx.doi.org/10.1007/s000110050495] [PMID: 10563466]

[21] Rahbar, S.; Figarola, J.L. Novel inhibitors of advanced glycation endproducts. *Arch. Biochem. Biophys.*, **2003**, *419*(1), 63-79. [http://dx.doi.org/10.1016/j.abb.2003.08.009] [PMID: 14568010]

[22] Boel, E.; Selmer, J.; Flodgaard, H.J.; Jensen, T. Diabetic late complications: Will aldose reductase inhibitors or inhibitors of advanced glycosylation endproduct formation hold promise? *J. Diabetes Complications,* **1995**, *9*(2), 104-129. [http://dx.doi.org/10.1016/1056-8727(94)00025-J] [PMID: 7599349]

[23] John, L. Big problem for BioStratum. *Triangle Business J,* **2005**.

[24] Malik, N.S.; Meek, K.M. The inhibition of sugar-induced structural alterations in collagen by aspirin and other compounds. *Biochem. Biophys. Res. Commun.*, **1994**, *199*(2), 683-686. [http://dx.doi.org/10.1006/bbrc.1994.1282] [PMID: 8135810]

[25] van Boekel, M.A.M.; van den Bergh, P.J.P.C.; Hoenders, H.J. Glycation of human serum albumin: inhibition by Diclofenac. *Biochim. Biophys. Acta Protein Struct. Mol. Enzymol.*, **1992**, *1120*(2), 201-204. [http://dx.doi.org/10.1016/0167-4838(92)90270-N] [PMID: 1562587]

[26] Tarasuntisuk, S.; Patipong, T.; Hibino, T.; Waditee-Sirisattha, R.; Kageyama, H. Inhibitory effects of mycosporine-2-glycine isolated from a halotolerant cyanobacterium on protein glycation and collagenase activity. *Lett. Appl. Microbiol.*, **2018**, *67*(3), 314-320. [http://dx.doi.org/10.1111/lam.13041] [PMID: 29947423]

[27] Orfanoudaki, M.; Hartmann, A.; Alilou, M.; Gelbrich, T.; Planchenault, P.; Derbré, S.; Schinkovitz, A.; Richomme, P.; Hensel, A.; Ganzera, M. Absolute configuration of mycosporine-like amino acids, their wound healing properties and *in vitro* anti-aging effects. *Mar. Drugs,* **2019**, *18*(1), 35. [http://dx.doi.org/10.3390/md18010035] [PMID: 31906052]

[28] Waditee-Sirisattha, R.; Kageyama, H. Protective effects of mycosporine-like amino acid-containing emulsions on UV-treated mouse ear tissue from the viewpoints of antioxidation and antiglycation. *J. Photochem. Photobiol. B,* **2021**, *223*, 112296. [http://dx.doi.org/10.1016/j.jphotobiol.2021.112296] [PMID: 34450363]

CHAPTER 9

Biological Activities of MAAs and their Applications 5: Inhibition of Collagenase Activity

Abstract: Enzymes involved in the degradation of the extracellular matrix (ECM) are deeply involved in skin aging. Compounds that suppress the degradation of collagen and elastin, constituents of the ECM, are of significant value to the cosmetics field. So far, more than 10 types of MAAs have been reported to inhibit the activity of collagenase, which belong to the family of matrix metalloproteinases. It has been suggested that the metal-chelating activity of MAAs is involved in these mechanisms of action. However, MAAs have not been reported to have an inhibitory activity on elastase. This chapter briefly summarizes these observations.

Keywords: Collagenase, Elastase, Extracellular matrix, Matrix metalloproteinase, Metal chelating activity.

INTRODUCTION

Collagen is a fibrous protein that constitutes various tissues including skin, bone, and blood vessels in vertebrates (Fig. **1**). Collagen accounts for 30% of proteins that make up humans. Collagen is produced by fibroblasts within the dermis layer. It has high strength and is involved in the maintenance of structure in the skin tissue, which greatly affects the elasticity and firmness of the skin. Because collagen depletion and denaturation cause skin aging, prevention of collagen depletion leads to skin anti-aging. Therefore, compounds with an inhibitory effect on collagenase, which is a collagen-degrading enzyme, are useful in the cosmetics field.

Collagenase, which is an endopeptidase, belongs to the family of matrix metalloproteinases (MMPs). By degrading collagen and elastin, MMPs play an important role in a variety of biological processes, including tissue homeostasis and post-wound repair. However, MMPs are activated with aging and change the composition of collagen and elastin in the extracellular matrix (ECM), resulting in wrinkles and sagging of the skin [1]. Several types of MMPs are expressed in mammalian skin [2, 3]. Mammalian collagenases are secreted by keratinocytes and skin fibroblasts in response to UV irradiation, oxidative stress, and cytokine

Hakuto Kageyama

stimulation. Repeated induction of these collagen-degrading enzymes over a long accelerates skin aging. Bacterial collagenase is known to be one of the factors of bacterial virulence. It destroys the extracellular structure by attacking the collagen helix and is involved in the pathogenic process of some bacteria, such as *Clostridium* [3, 4].

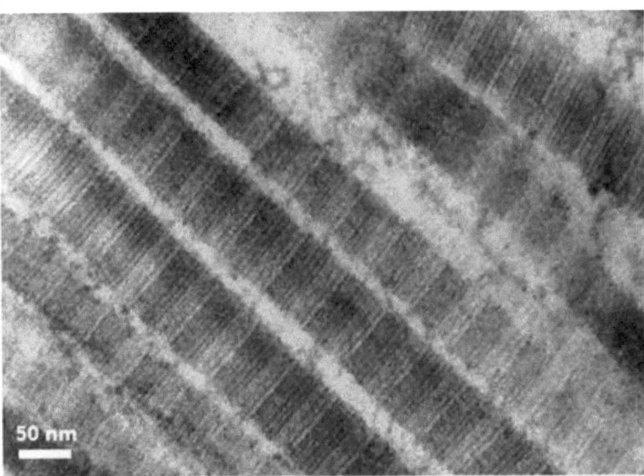

Fig. (1). Transmission electron microscope image of fibrous collagen (type I collagen).

INHIBITORY EFFECTS OF MAAS ON COLLAGENASE ACTIVITIES

So far, 14 types of MAAs have been reported to have an inhibitory effect on collagenase activity in *in vitro* experiments (Table **1**). Hartmann *et al.* reported the collagenase-inhibiting activity of shinorine, porphyra-334, and palythine isolated from red algae [5]. Another research group showed the collagenase-inhibiting activity of mycosporine-2-glycine isolated from halotolerant cyanobacterium *Halothece* sp. PCC7418 [6]. Interestingly, in that report, no collagenase inhibitory activity was detected in the mixture of shinorine and porphyra-334 [6]. Orfanoudaki *et al.* also reported the collagenase-inhibiting activity of 13 MAAs, including shinorine and porphyra-334 [7]. Another report found that the amount of MMPs accumulated after applying UV irradiation to mouse skin tissue was suppressed by the application of MAAs [8]. The mechanism of collagenase inhibition by MAAs is not fully understood. As outlined in Chapter 10, MAAs may be a chelating agent for metal ions [9], which may lead to the inhibition of metal-requiring collagenase.

In addition to collagenase, elastase, a member of the chymotrypsin-type serine protease family, is an enzyme involved in ECM degradation [10]. Degradation of elastin leads to a decrease in skin elasticity. So far, no results have reported that MAAs showed inhibition of elastase activity. In a preliminary study using elastase from porcine pancreas by a research group in Japan, the purified MAAs (M2G,

P334, and SHI) tested showed no inhibitory activity [3]. Unlike collagenase, it should be noted that elastase does not require metal ions to exhibit protease activity.

Table 1. MAAs reported to have an inhibitory effect on collagenase activity.

MAA	Collagenase (Substrate)	Activity (IC_{50})
Shinorine [5]	Collagenase type V from *Clostridium histolyticum* (MMP-2 substrate SCP0192)	104 μM
Porphyra-334 [5]	Collagenase type V from *C. histolyticum* (MMP-2 substrate SCP0192)	106 μM
Palythine [5]	Collagenase type V from *C. histolyticum* (MMP-2 substrate SCP0192)	159 μM
Mycosporine-2-glycine [6]	Collagenase type IV from *C. histolyticum* (4-phenylazobenzyloxycarbonyl-Pro-Leu-Gly-Pro-D-Arg-OH)	470 μM
Mycosporine-methylamine-threonine [7]	Collagenase type V from *C. histolyticum* (MMP-2 substrate SCP0192)	251 μM
Mycosporine-alanine-glycine [7]	Collagenase type V from *C. histolyticum* (MMP-2 substrate SCP0192)	158 μM
Aplysiapalythine A [7]	Collagenase type V from *C. histolyticum* (MMP-2 substrate SCP0192)	81 μM
Asterina-330 [7]	Collagenase type V from *C. histolyticum* (MMP-2 substrate SCP0192)	71 μM
Bostrychine B [7]	Collagenase type V from *C. histolyticum* (MMP-2 substrate SCP0192)	105 μM
Bostrychine C [7]	Collagenase type V from *C. histolyticum* (MMP-2 substrate SCP0192)	58 μM
Bostrychine D [7]	Collagenase type V from *C. histolyticum* (MMP-2 substrate SCP0192)	118 μM
Bostrychine E [7]	Collagenase type V from *C. histolyticum* (MMP-2 substrate SCP0192)	163 μM

(Table 1) cont.....

MAA	Collagenase (Substrate)	Activity (IC_{50})
Bostrychine F [7]	Collagenase type V from *C. histolyticum* (MMP-2 substrate SCP0192)	90 μM
Mycosporine-glycine [7]	Collagenase type V from *C. histolyticum* (MMP-2 substrate SCP0192)	81 μM

MMP-2 substrate SCP0192: MCA-Pro-Leu-Ala-Nva-DNP-Dap-Ala-Arg-NH_2.

CONCLUDING REMARKS

This chapter outlined the ability of MAAs to inhibit collagenase activity, which is a characteristic that can contribute to the cosmetics field. So far, several research groups have confirmed these activities. Research on the relationship between the degree of activity and the molecular structure of MAAs will be an interesting subject in the future.

LIST OF ABBREVIATIONS

ECM extracellular matrix

MAA mycosporine-like amino acid

MMP matrix metalloproteinase

CONSENT FOR PUBLICATION

Not applicable.

CONFLICT OF INTEREST

The author declares no conflict of interest, financial or otherwise.

ACKNOWLEDGEMENTS

Declared none.

REFERENCES

[1] Pittayapruek, P.; Meephansan, J.; Prapapan, O.; Komine, M.; Ohtsuki, M. Role of matrix metalloproteinases in photoaging and photocarcinogenesis. *Int. J. Mol. Sci.,* **2016**, *17*(6), 868. [http://dx.doi.org/10.3390/ijms17060868] [PMID: 27271600]

[2] Brennan, M.; Bhatti, H.; Nerusu, K.C.; Bhagavathula, N.; Kang, S.; Fisher, G.J.; Varani, J.; Voorhees, J.J. Matrix metalloproteinase-1 is the major collagenolytic enzyme responsible for collagen damage in UV-irradiated human skin. *Photochem. Photobiol.,* **2003**, *78*(1), 43-48. [http://dx.doi.org/10.1562/0031-8655(2003)078<0043:MMITMC>2.0.CO;2] [PMID: 12929747]

[3] Kageyama, H.; Waditee-Sirisattha, R. Antioxidative, anti-inflammatory, and anti-aging properties of mycosporine-like amino acids: Molecular and cellular mechanisms in the protection of skin-aging. *Mar. Drugs,* **2019**, *17*(4), 222.

[http://dx.doi.org/10.3390/md17040222] [PMID: 31013795]

[4] Duarte, A.S.; Correia, A.; Esteves, A.C. Bacterial collagenases – A review. *Crit. Rev. Microbiol.,* **2016**, *42*(1), 106-126.
[http://dx.doi.org/10.3109/1040841X.2014.904270] [PMID: 24754251]

[5] Hartmann, A.; Gostner, J.; Fuchs, J.; Chaita, E.; Aligiannis, N.; Skaltsounis, L.; Ganzera, M. Inhibition of collagenase by mycosporine-like amino acids from marine sources. *Planta Med.,* **2015**, *81*(10), 813-820.
[http://dx.doi.org/10.1055/s-0035-1546105] [PMID: 26039265]

[6] Tarasuntisuk, S.; Patipong, T.; Hibino, T.; Waditee-Sirisattha, R.; Kageyama, H. Inhibitory effects of mycosporine-2-glycine isolated from a halotolerant cyanobacterium on protein glycation and collagenase activity. *Lett. Appl. Microbiol.,* **2018**, *67*(3), 314-320.
[http://dx.doi.org/10.1111/lam.13041] [PMID: 29947423]

[7] Orfanoudaki, M.; Hartmann, A.; Alilou, M.; Gelbrich, T.; Planchenault, P.; Derbré, S.; Schinkovitz, A.; Richomme, P.; Hensel, A.; Ganzera, M. Absolute configuration of mycosporine-like amino acids, their wound healing properties and *in vitro* anti-aging effects. *Mar. Drugs,* **2019**, *18*(1), 35.
[http://dx.doi.org/10.3390/md18010035] [PMID: 31906052]

[8] Ying, R.; Zhang, Z.; Zhu, H.; Li, B.; Hou, H. The protective effect of mycosporine-like amino acids (MAAs) from *Porphyra yezoensis* in a mouse model of UV irradiation-induced photoaging. *Mar. Drugs,* **2019**, *17*(8), 470.
[http://dx.doi.org/10.3390/md17080470] [PMID: 31416181]

[9] Varnali, T.; Bozoflu, M.; Şengönül, H.; Kurt, S.İ. Potential metal chelating ability of mycosporine-like amino acids: a computational research. *Chem. Zvesti,* **2022**, *76*(4), 2279-2291.
[http://dx.doi.org/10.1007/s11696-021-02014-x]

[10] Thring, T.S.A.; Hili, P.; Naughton, D.P. Anti-collagenase, anti-elastase and anti-oxidant activities of extracts from 21 plants. *BMC Complement. Altern. Med.,* **2009**, *9*(1), 27.
[http://dx.doi.org/10.1186/1472-6882-9-27] [PMID: 19653897]

CHAPTER 10

Biological Activities of MAAs and their Applications 6: Metal Chelating Abilities

Abstract: As mentioned in Chapters 8 and 9, the useful functions of MAAs, such as the anti-glycative property and collagenase inhibitory activity, might be associated with their metal chelating activity. Although there are few reports on the metal-chelating activity of MAAs, a chelating model of MAAs and metal ions has recently been proposed. This chapter briefly summarizes these observations.

Keywords: Chelating, Euhalothece-362, Mycosporine-2-glycine.

INTRODUCTION

A compound in which a ligand (*i.e.*, a molecule or ion) is bonded to a metal ion is called a complex. If the ligand has multiple atoms that coordinate with the metal ion, a ring structure containing the metal ion is formed. This compound is called a chelate compound. The etymology of chelate is 'chēlē', which means "crab scissors" in Greek. A chelating agent binds to a metal ion in solution and reduces the activity of the metal ion. Chelating agents have various applications. For example, they are added to shampoos and laundry detergents to prevent the salt formation of anionic surfactants and maintain detergency. In agriculture, chelate compounds are used as water-soluble metal salt fertilizers due to their high solubility in water.

Fig. (**1A**) shows the structure of a chelate compound of a divalent nickel ion and ethylenediaminetetraacetic acid (EDTA), which is a major chelating agent. The arrows in the figure indicate coordinate bonds, indicating that lone electron pairs are provided by oxygen and nitrogen atoms*. Amino acids with an amino group ($-NH_2$) and a carboxyl group ($-COOH$) are also known to act as chelating agents. Fig. (**1B**) shows the structure of a chelate compound of a divalent calcium ion and glycine. Because MAAs contain amino acids in their molecular structure, they have the potential to function as chelating agents.

Hakuto Kageyama

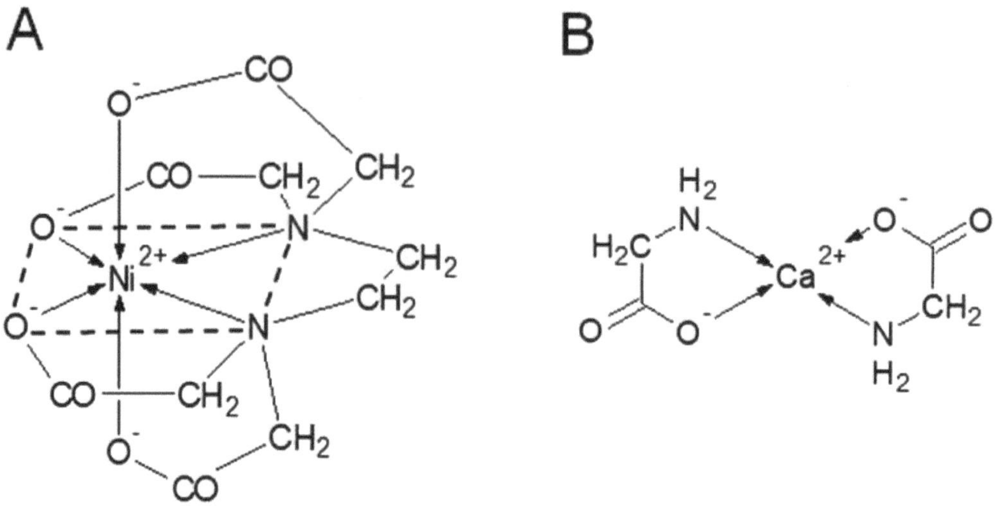

Fig. (1). Representative chelate compounds **A.** Chelate compound of Ni2+ and EDTA (4-) ion. **B.** Chelate compound of Ca^{2+} and glycine.

POTENTIAL METAL CHELATING ACTIVITIES OF MAAS

In 2006, Volkmann *et al.* suggested that euhalothece-362 (Fig. **2**), an MAA derived from the halophilic cyanobacterium *Euhalothece* sp. strain LK-1 may act as a chelating agent [1]. At that time, it was presumed that the four hydroxy groups (−OH) present in the structure of enhalothece-362 and the carboxy group of the alanine residue substituted at the C3 position were involved in chelation.

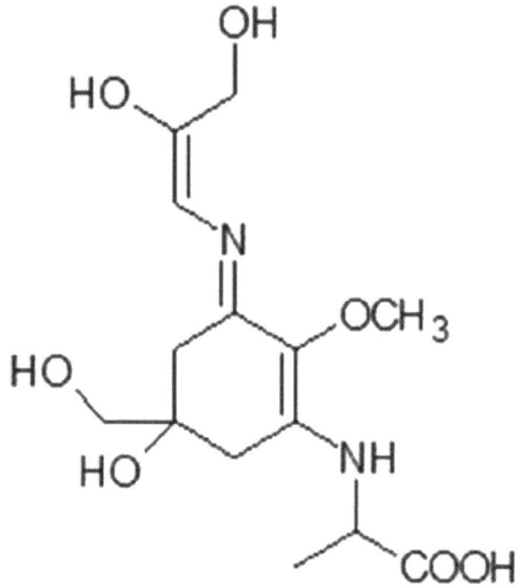

Fig. 2. Molecular structure of euhalothece-362.

Later, Varnali *et al.* proposed a model structure for chelated compounds of metal ions and MAAs [2]. Fig. (**3**) shows two examples of the chelation of mycosporine-glycine and calcium ions. In the first model, the oxygen atom of the methoxy group ($-OCH_3$) at the C2 position and the nitrogen atom and oxygen atom of the carboxy group of the glycine residue substituted at the C3 position of mycosporine-glycine are bound to the calcium ion (Fig. **3A**). In the second model, the two oxygen atoms of the carboxy group of glycine and the oxygen atoms of the hydroxy group and hydroxymethyl group ($-CH_2$-OH) at the C5 position bind to the calcium ion (Fig. **3B**).

Fig. (3). Two chelation models of mycosporine-glycine and Ca^{2+}.

In addition, in the disubstituted MAs shinorine, a model has been proposed in which the hydroxy group of the serine residues substituted at the C1 position is bound to the metal ion (Fig. **4**).

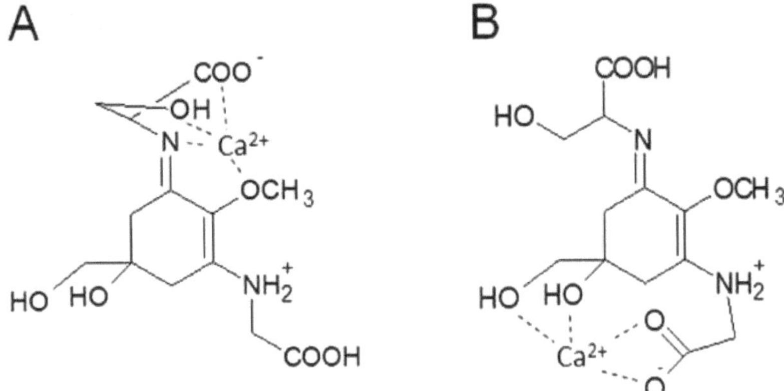

Fig. (4). Two chelation models of shinorine and Ca^{2+}.

Another study reported that mycosporine-2-glycine has a chelating ability for divalent iron ions. In that report, mycosporine-2-glycine purified from the halotolerant cyanobacterium *Halothece* sp. PCC7418 exhibited the metal chelating activity in a concentration-dependent manner, whereas a mixture of porphyra-334 and shinorine did not have a remarkable metal chelating activity [3]. Future analysis using other MAAs is awaited.

CONCLUDING REMARKS

This chapter summarized the experimental results and proposed models of the metal chelating activity of MAAs. Further investigation is needed on this important property of MAAs.

CONSENT FOR PUBLICATION

Not applicable.

CONFLICT OF INTEREST

The author declares no conflict of interest, financial or otherwise.

ACKNOWLEDGEMENTS

Declared none.

REFERENCES

[1] Volkmann, M.; Gorbushina, A.A.; Kedar, L.; Oren, A. Structure of euhalothece-362, a novel red-shifted mycosporine-like amino acid, from a halophilic cyanobacterium (Euhalothece sp.). *FEMS Microbiol. Lett.,* **2006**, *258*(1), 50-54. [http://dx.doi.org/10.1111/j.1574-6968.2006.00203.x] [PMID: 16630254]

[2] Varnali, T.; Bozoflu, M.; Şengönül, H.; Kurt, S.İ. Potential metal chelating ability of mycosporine-like amino acids: a computational research. *Chem. Zvesti,* **2022**, *76*(4), 2279-2291. [http://dx.doi.org/10.1007/s11696-021-02014-x]

[3] Tarasuntisuk, S.; Palaga, T.; Kageyama, H.; Waditee-Sirisattha, R. Mycosporine-2-glycine exerts anti-inflammatory and antioxidant effects in lipopolysaccharide (LPS)-stimulated RAW 264.7 macrophages. *Arch. Biochem. Biophys.,* **2019**, *662*, 33-39. [http://dx.doi.org/10.1016/j.abb.2018.11.026] [PMID: 30502329]

CHAPTER 11

Biological Activities of MAAs and their Applications 7: DNA Protective Property, Wound Healing Effects, Anti-Cancer Effects, And Applications in Horticulture and as a Film Material

Abstract: This chapter describes the research findings on the DNA protection, wound healing, and anti-cancer effects of MAAs and their use in horticulture and as a raw material for film-type UV-blocking materials. These applications are additional to the use of MAAs as sunscreen agents and pharmaceuticals applied to humans.

Keywords: Chelating, DNA protection, Euhalothece-362, Mycosporine-2-glycine, Wound healing.

INTRODUCTION

DNA Protective Properties of MAAS

It is known that reactive oxygen species (ROS) and oxidative stress generated by UV irradiation cause oxidative DNA damage. It has been reported that MAAs can reduce the DNA damage caused by these external stresses. It was mentioned in Chapter 5 that Helioguard 365 relieved DNA damage from UV irradiation in human fibroblasts. In addition, DNA damage caused by oxidative stress treatment with hydrogen peroxide in human malignant melanoma A375 cells was suppressed by the addition of mycosporine-2-glycine to the medium [1]. It is thought that MAAs, which have an antioxidant capacity, protect DNA by scavenging ROS.

UV is absorbed by the double bonds in the pyrimidine ring of thymine and cytosine, which are the constituent bases of DNA. As a result, the double bond is cleaved and can react with an adjacent base. A polymer of adjacent thymines and cytosines is called a pyrimidine dimer. Pyrimidine dimers inhibit DNA replication

Hakuto Kageyama

and transcription, causing various disorders such as cell death and mutation. It has been reported that red algae extracts containing palythine, shinorine, and porphyra-334 suppressed the production of pyrimidine dimers [2].

WOUND HEALING EFFECTS OF MAAS

MAAs might have a wound healing effect. Orfanoudaki *et al.* showed that shinorine, porphyra-334, mycosporine-alanine-glycine, and bostrychine-B showed the effect of closing scratches made on monolayer cultures of human keratinocyte cells by scratch assay* [3]. These MAAs are thought to have the effect of promoting the proliferation and migration of human keratinocyte cells, but the mechanism of action is still unknown.

* A method that scratches a monolayer culture on a plate and investigates the process in which the area is closed by cell proliferation or migration.

ANTI-CANCER EFFECTS OF MAAS

It has been reported that MAAs have an inhibitory effect on the growth of cancer cells. For example, Yuan *et al.* showed that an MAA-containing extract prepared from the red alga *Palmaria palmata* inhibited the growth of B16-F1 mouse skin melanoma cell line [4]. The extract prepared from red algae exposed to the high UV irradiation environment (grade II) had a higher growth inhibitory effect than that of the low UV irradiation environment (grade I). Because grade II contained usujirene in addition to palythine, palythinol, shinorine, asterina-330, and porphyra-334 contained in grade I, it is possible that usujirene contributed to growth inhibition. In another report, extracts prepared from wild- and cultivated-red algae showed a growth inhibitory effect on human HeLa adenocarcinoma cervical cell line and U-937 histiocytic lymphoma cell line *in vitro* [5]. These antiproliferative activities were thought to be triggered by the induction of apoptosis. Other reports also suggested that MAAs induce apoptosis. Kim *et al.* investigated the effects of an extract of the red alga *Porphyra yezoensis* containing porphyra-334 and shinorine on human keratinized HaCaT cells treated with UV-B. As a result, UV-B-irradiated cells were protected by promoting cell proliferation through the activation of the JNK and ERK signaling pathways and inducing the apoptosis of damaged cells [6].

APPLICATION OF MAAS TO HORTICULTURE

MAAs may be used to combat sunscald in horticultural crops. Although it is common to physically block direct sunlight to prevent sunscald, it may be possible to chemically protect crops by applying an MAA-containing emulsion to crops. From this perspective, Pedrosa *et al.* prepared a stable emulsion containing carnauba wax in an ammonium aqueous solution to which Helioguard 365 was added to enhance absorption in the UV-B region (280–300 nm) [7]. However, the actual application to crops has not been examined, so future development is expected.

APPLICATION OF MAAS AS FILM MATERIALS

MAAs are useful as a raw material for film-type UV-blocking materials. For example, Fernandes *et al.* prepared and evaluated MAA-containing (mycosporine-glycine, shinorine, or porphyra-334) films using chitosan as a medium [8]. These films were prepared by forming an amide bond between the carboxy group of MAAs and the amino group of chitosan. As a result, the film effectively absorbed the UV-A and UV-B regions and showed light and heat resistance. The UV absorption pattern was different depending on the MAA species used. In addition, the biocompatibility of these films was confirmed using L-929 murine fibroblasts. Because substances other than chitosan can be used as the medium, it is possible that various functional films can be developed using MAAs.

CONCLUDING REMARKS

This chapter outlined the research findings on DNA protection, wound healing, and anti-cancer effects as potential useful functions of MAAs. Regarding these effects, it is necessary to perform more detailed studies in the future. In addition, application examples of MAAs as a film material and an anti-burning agent were described. By utilizing the properties of MAAs such as UV absorption and antioxidant activity, it is possible that MAAs can be applied as various materials in the future.

CONSENT FOR PUBLICATION

Not applicable.

CONFLICT OF INTEREST

The author declares no conflict of interest, financial or otherwise.

ACKNOWLEDGEMENTS

Declared none.

REFERENCES

[1] Cheewinthamrongrod, V.; Kageyama, H.; Palaga, T.; Takabe, T.; Waditee-Sirisattha, R. DNA damage protecting and free radical scavenging properties of mycosporine-2-glycine from the Dead Sea

cyanobacterium in A375 human melanoma cell lines. *J. Photochem. Photobiol. B,* **2016**, *164*, 289-295. [http://dx.doi.org/10.1016/j.jphotobiol.2016.09.037] [PMID: 27718421]

[2] Misonou, T.; Saitoh, J.; Oshiba, S.; Tokitomo, Y.; Maegawa, M.; Inoue, Y.; Hori, H.; Sakurai, T. UV-absorbing substance in the red alga *Porphyra yezoensis* (Bangiales, Rhodophyta) block thymine photodimer production. *Mar. Biotechnol. (NY),* **2003**, *5*(2), 194-200. [http://dx.doi.org/10.1007/s10126-002-0065-2] [PMID: 12876656]

[3] Orfanoudaki, M.; Hartmann, A.; Alilou, M.; Gelbrich, T.; Planchenault, P.; Derbré, S.; Schinkovitz, A.; Richomme, P.; Hensel, A.; Ganzera, M. Absolute configuration of mycosporine-like amino acids, their wound healing properties and *in vitro* anti-aging effects. *Mar. Drugs,* **2019**, *18*(1), 35. [http://dx.doi.org/10.3390/md18010035] [PMID: 31906052]

[4] Yuan, Y.V.; Westcott, N.D.; Hu, C.; Kitts, D.D. Mycosporine-like amino acid composition of the edible red alga, *Palmaria palmata* (dulse) harvested from the west and east coasts of Grand Manan Island, New Brunswick. *Food Chem.,* **2009**, *112*(2), 321-328. [http://dx.doi.org/10.1016/j.foodchem.2008.05.066]

[5] Athukorala, Y.; Trang, S.; Kwok, C.; Yuan, Y. Antiproliferative and antioxidant activities and mycosporine-like amino acid profiles of wild-harvested and cultivated edible Canadian marine red macroalgae. *Molecules,* **2016**, *21*(1), 119. [http://dx.doi.org/10.3390/molecules21010119] [PMID: 26805798]

[6] Kim, S.; You, D.H.; Han, T.; Choi, E.M. Modulation of viability and apoptosis of UVB-exposed human keratinocyte HaCaT cells by aqueous methanol extract of laver (Porphyra yezoensis). *J. Photochem. Photobiol. B,* **2014**, *141*, 301-307. [http://dx.doi.org/10.1016/j.jphotobiol.2014.10.012] [PMID: 25463682]

[7] Pedrosa, V.M.; Sanches, A.G.; da Silva, M.B.; Gratao, P.L.; Isaac, V.L.; Gindri, M. Production of mycosporine-like amino acid (MAA)-loaded emulsions as chemical barriers to control sunscald in fruits and vegetables. *J. Sci. Food Agric.,* **2021**. [PMID: 34223643]

[8] Fernandes, S.C.M.; Alonso-Varona, A.; Palomares, T.; Zubillaga, V.; Labidi, J.; Bulone, V. Exploiting mycosporines as natural molecular sunscreens for the fabrication of UV-absorbing green materials. *ACS Appl. Mater. Interfaces,* **2015**, *7*(30), 16558-16564. [http://dx.doi.org/10.1021/acsami.5b04064] [PMID: 26168193]

Appendix A

Molecular structures, molecular weights, absorption maxima, Etinction coefficients of MAAs

MAA	Molecular weight (MW) Apsorption maxima (λ_{max}) Etinction coefficients (ε)
Precursor compound of MAAs	
4-Deoxygadusol [1] 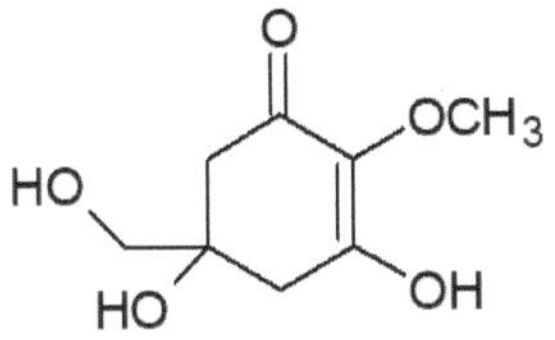	MW: 188 λ_{max} = 268 nm (pH = 2), 294 nm (pH =7) ε = ND
Monosubstituted-MAAs	
Mycosporine-glycine [2] 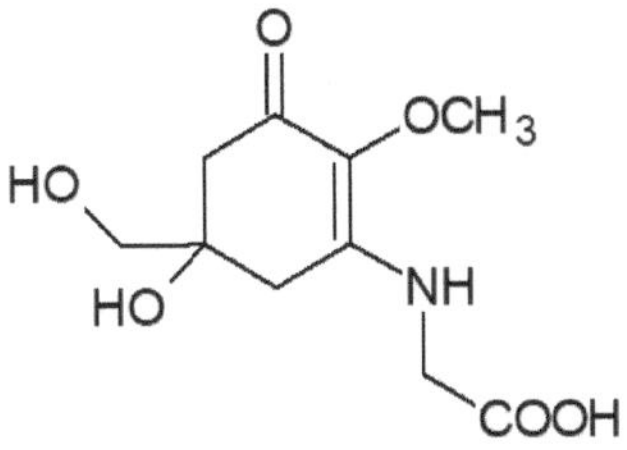	MW: 245 λ_{max} = 310 nm ε = ND
Mycosporine-taurine[3] 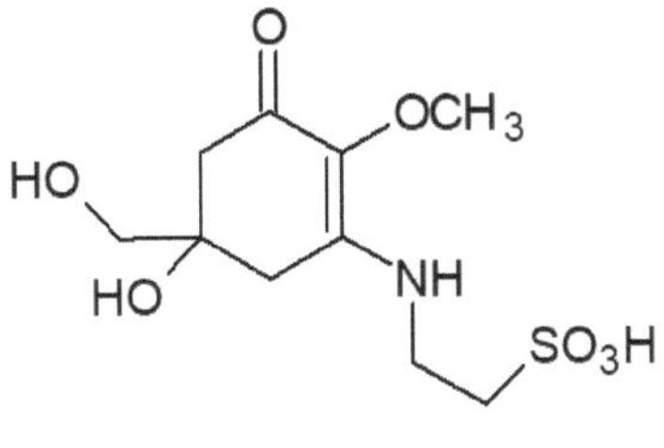	MW: 295 λ_{max} = 309 nm ε = ND

Hakuto Kageyama

Compound	Properties
Mycosporine-alanine [4, 5] 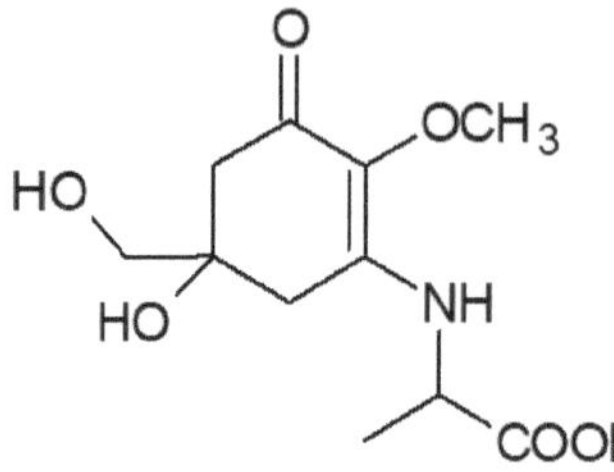 O, OCH3, HO, HO, NH, COOH	MW: 259 λ_{max} = 310 nm[4], 317 nm[5] ε = 640 M^{-1} cm^{-1}
Mycosporine-GABA[6] (Mycosporine-γ-aminobutyric acid) 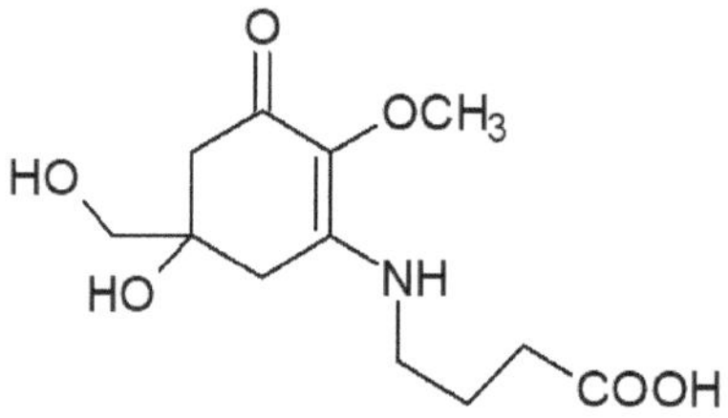 O, OCH3, HO, HO, NH, COOH	MW: 273 λ_{max} = 310 nm ε = 28,900 M^{-1} cm^{-1}
Mycosporine-serine[7] 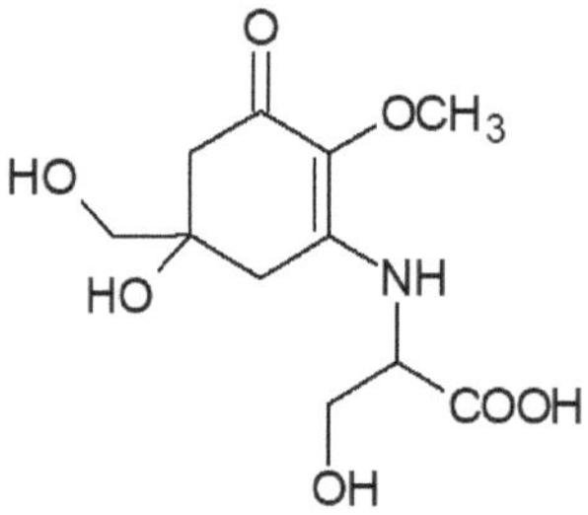 O, OCH3, HO, HO, NH, COOH, OH	MW: 275 λ_{max} = 310 nm ε = ND
Mycosporine-lysine [8] 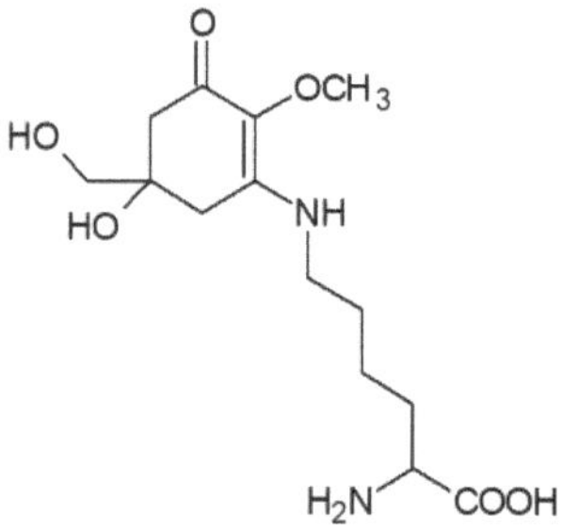 O, OCH3, HO, HO, NH, H2N, COOH	MW: 316 λ_{max} = 310 nm ε = ND

Compound	Properties
Mycosporine-ornithine [8] 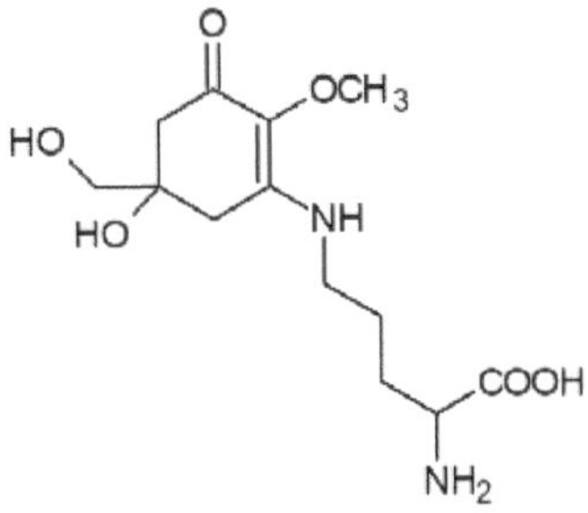	MW: 302 λ_{max} = 310 nm ε = ND
Mycosporine-ornithine (isomer) [9] 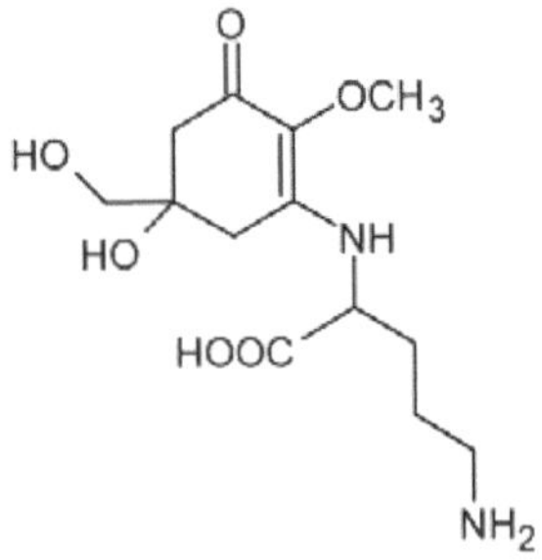	MW: 302 λ_{max} = ND ε = ND
Mycosporine-serinol [10] 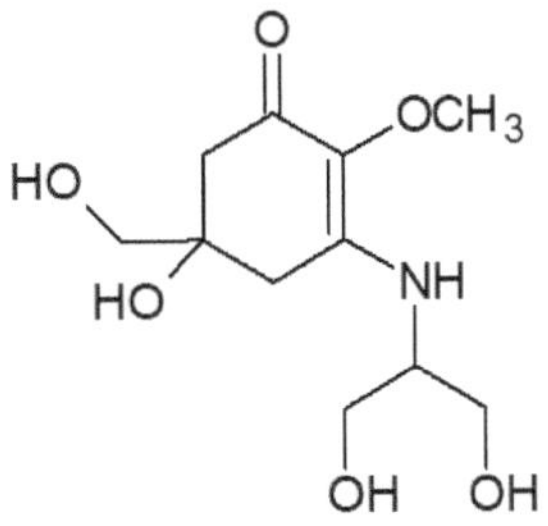	MW: 261 λmax = 310 nm ε = 27,270 M^{-1} cm^{-1}
Mycosporine-glutamic acid [11] 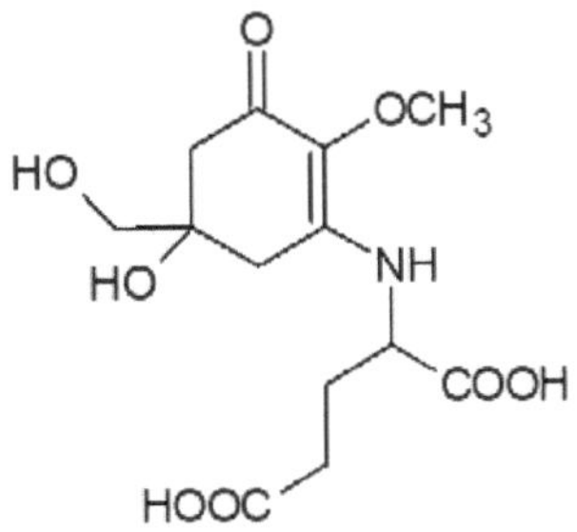	MW: 317 λ_{max} = 311 nm ε = 20,900 M^{-1} cm^{-1}

Mycosporine-glutamicol [12] 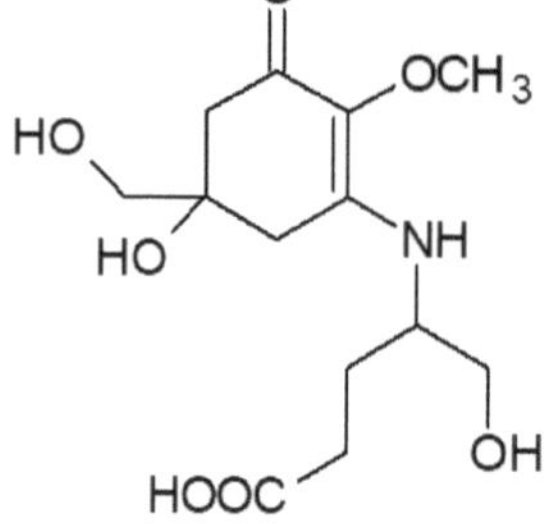	MW: 303 λ_{max} = 310 nm ε = ND
Mycosporine-glutamine [13] 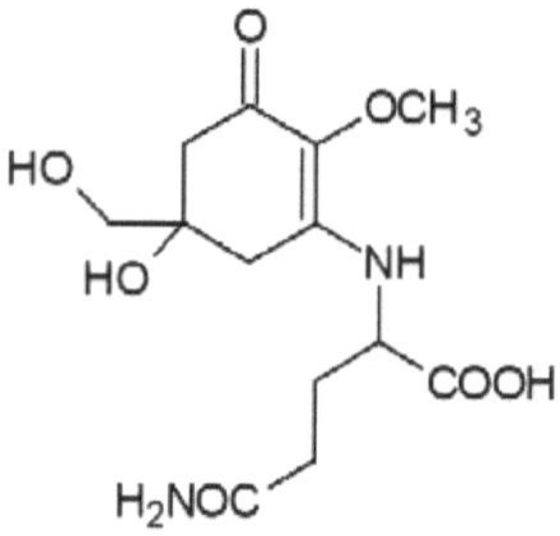	MW: 316 λ_{max} = 310 nm ε = ND
Mycosporine-glutaminol [14] 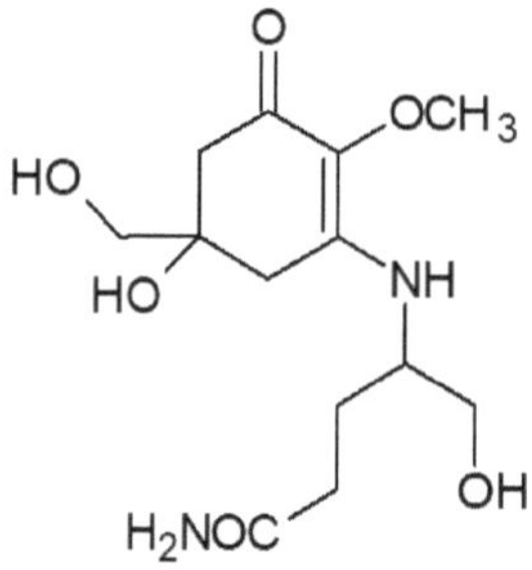	MW: 302 λ_{max} = ND ε = ND
Klebsormidin B [15] 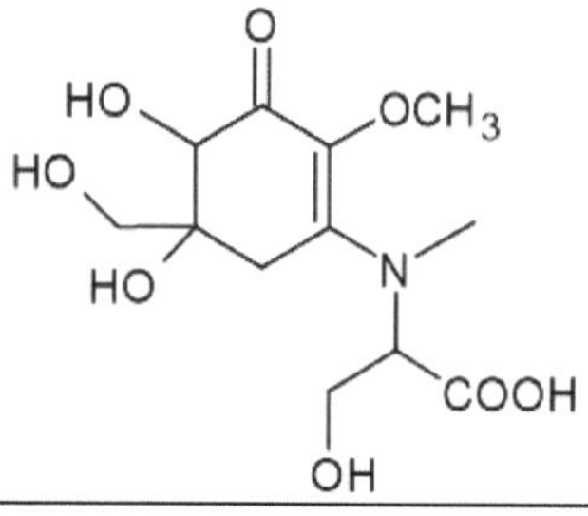	MW: 305 λ_{max} = 324 nm ε = ND

Disubstituted-MAAs	
Shinorine [1, 16] 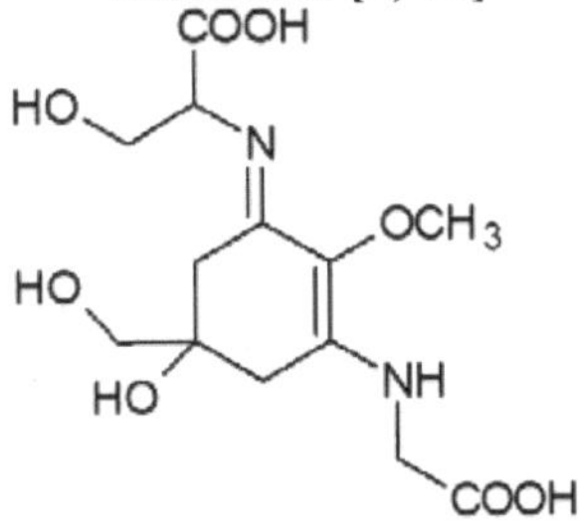	MW: 332 λ_{max} = 333-334 nm ε = 44,700 M^{-1} cm^{-1}
Porphyra-334 [17] 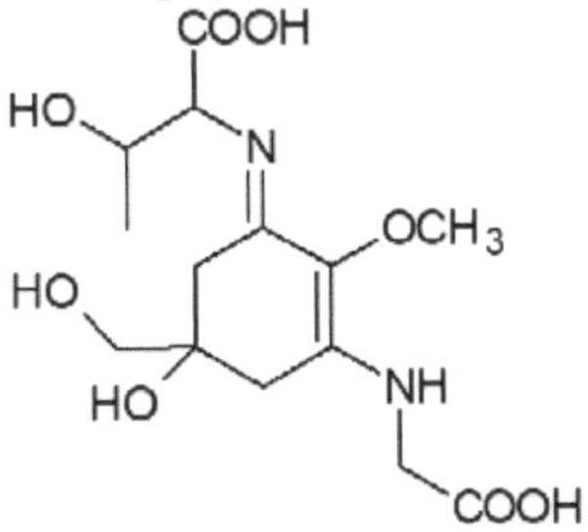	MW: 346 λ_{max} = 334 nm ε = 42,300 M^{-1} cm^{-1}
Mycosporine-2-glycine [18] 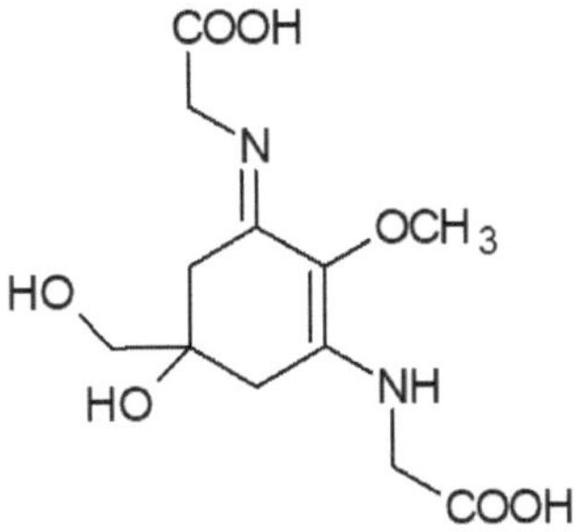	MW: 302 λ_{max} = 330-332 nm ε = ND
Asterina-330 [19] 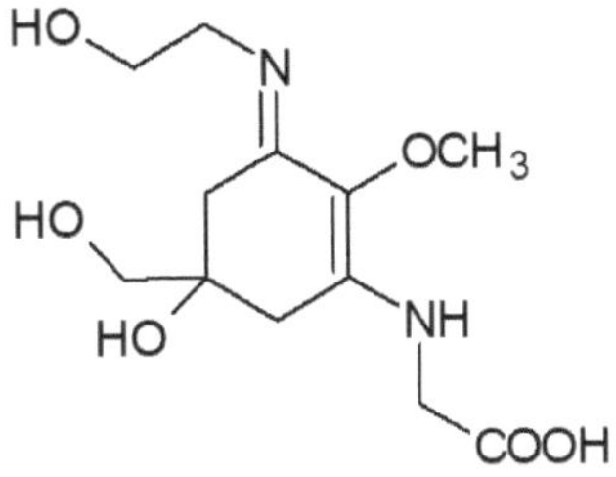	MW: 288 λ_{max} = 330 nm ε = 43,800 M^{-1} cm^{-1}

Compound	Properties
Mycosporine-glycine-glutamic acid [20] 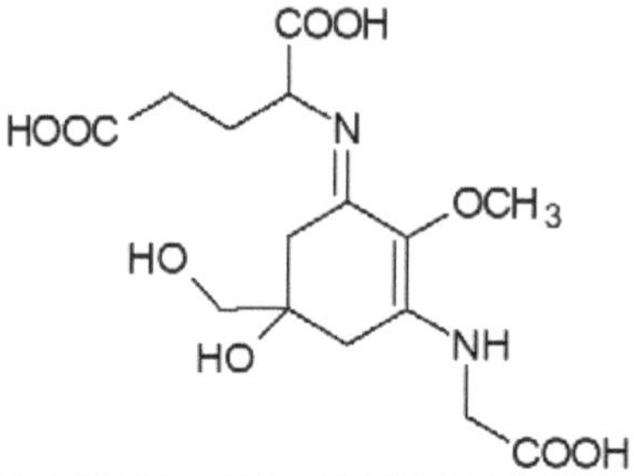	MW: 374 λ_{max} = 330 nm ε = 43,900 M^{-1} cm^{-1}
Mycosporine-glycine-aspartic acid [21] 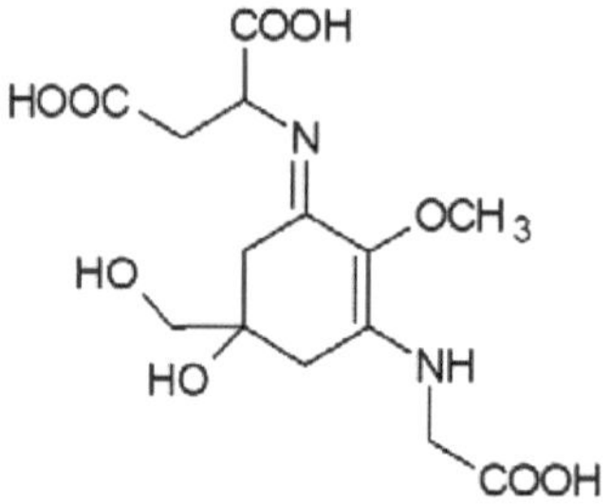	MW: 360 λ_{max} = 332-334 nm ε = ND
Mycosporine-glycine-valine [22] 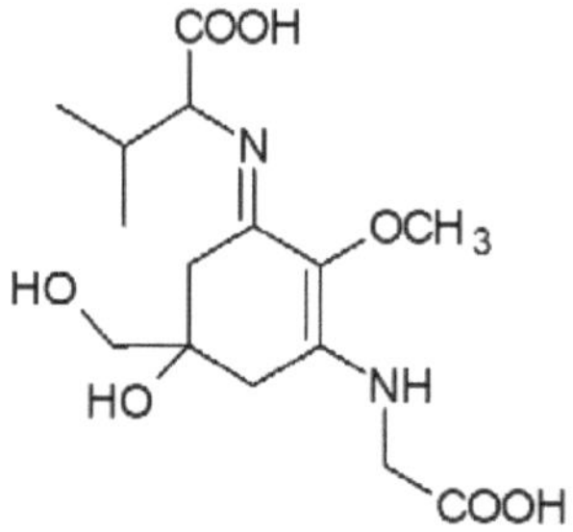	MW: 344 λ_{max} = 335 nm ε = ND
Mycosporine-glycine-alanine [23] 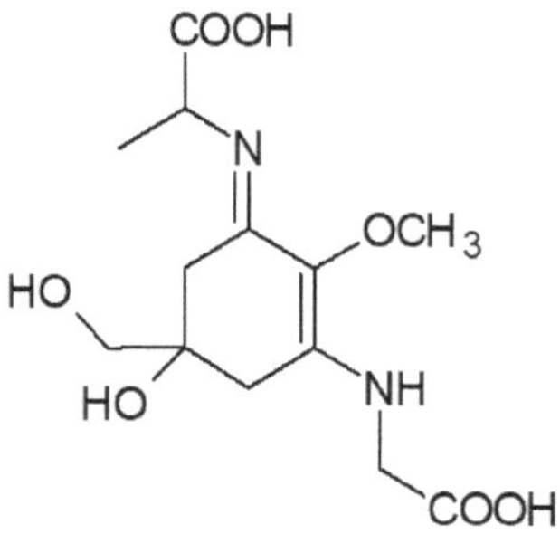	MW: 316 λ_{max} = 333 nm ε = ND

Compound	Properties
Mycosporine-glycine-arginine [24] 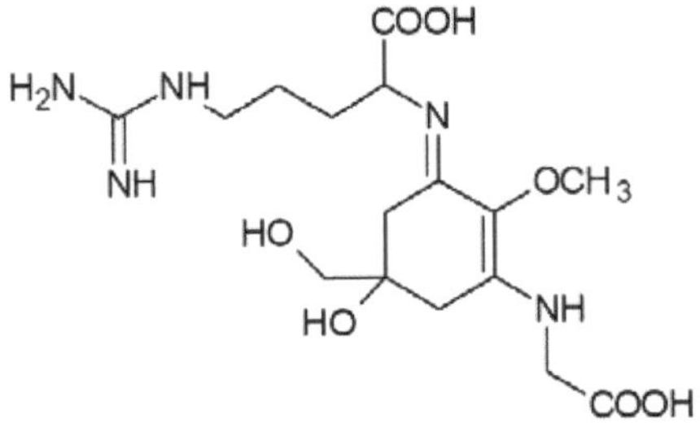	MW: 401 λ_{max} = 335 nm ε = ND
Mycosporine-glycine-cysteine [24] 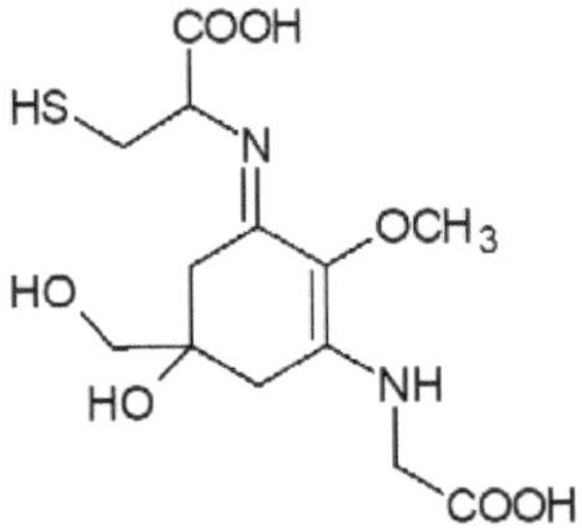	MW: 348 λ_{max} = 335 nm ε = ND
Usujirene [25] 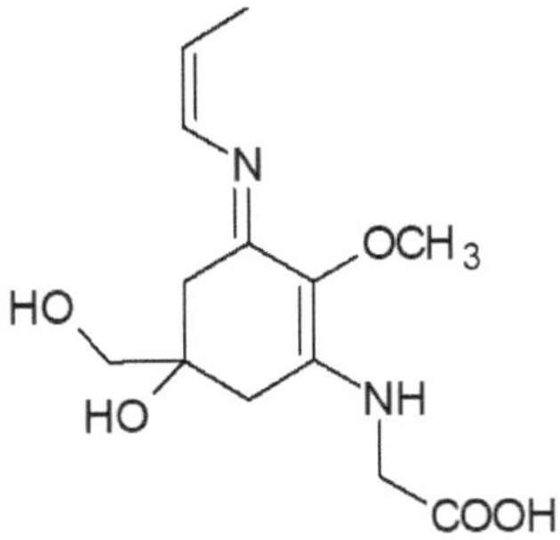	MW: 285 λ_{max} = 357 nm ε = ND
Palythene [26] 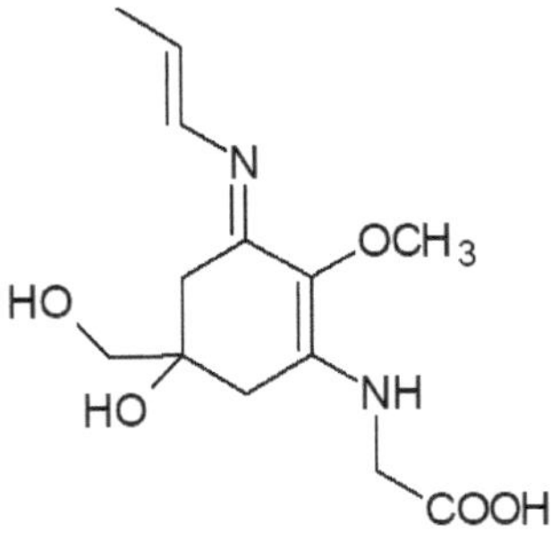	MW: 285 λ_{max} = 360 nm ε = 50,000 M^{-1} cm^{-1}

Palythenic acid [27] 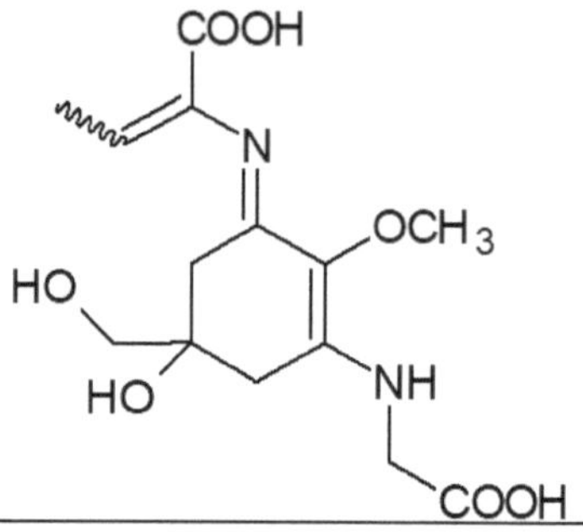 COOH N OCH 3 HO HO NH COOH	MW: 328 λ_{max} = 337 nm ε = 29,200 M^{-1} cm^{-1}
Palythinol [26] 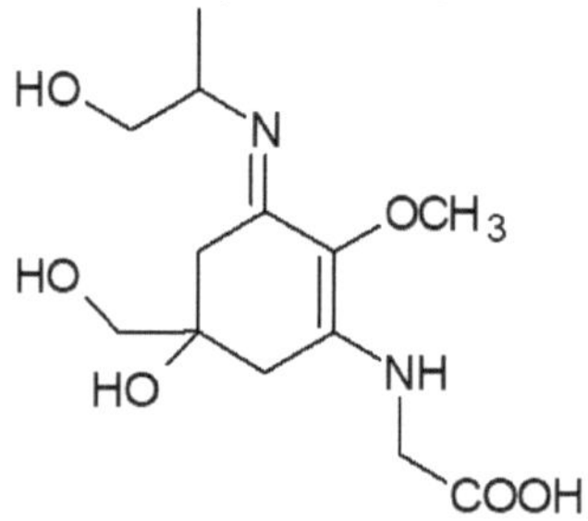 HO N OCH 3 HO HO NH COOH	MW: 302 λ_{max} = 332 nm ε = 43,500 M^{-1} cm^{-1}
Aplysiapalythine A [28] 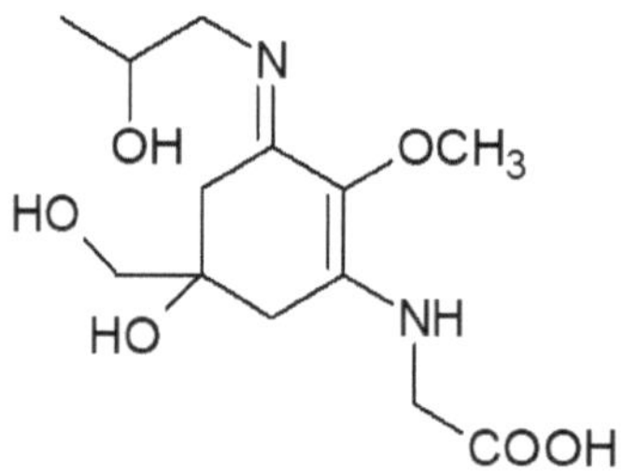 N OH OCH_3 HO HO NH COOH	MW: 302 λ_{max} = 332 nm ε = ND
Aplysiapalythine B [28] 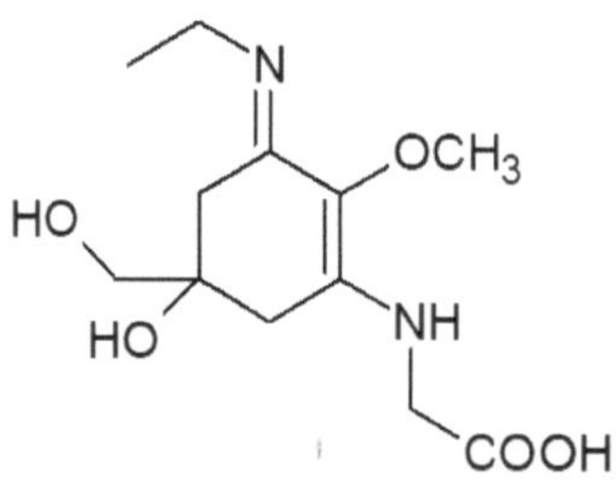 N OCH_3 HO HO NH COOH	MW: 272 λ_{max} = 332 nm ε = ND

Compound	Properties
Aplysiapalythine C [28] 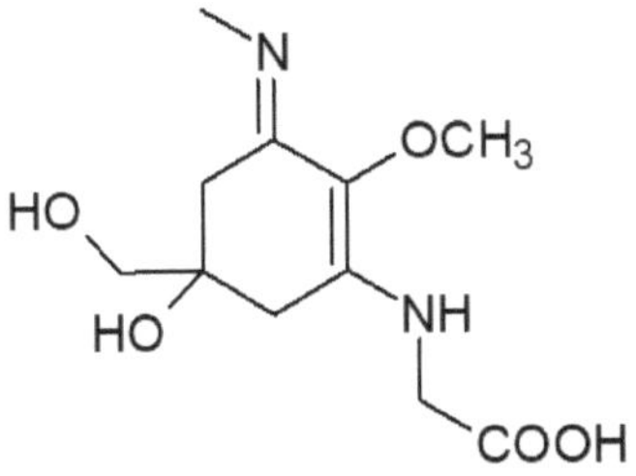	MW: 258 λ_{max} = 330 nm ε = ND
Aplysiapalythine D [29] 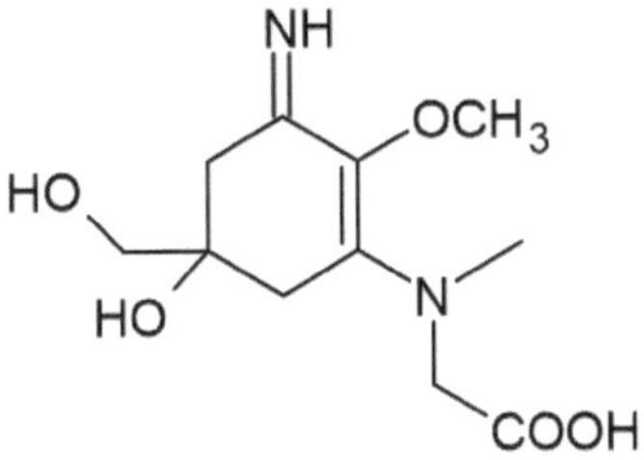	MW: 258 λ_{max} = 334 nm ε = ND
Coelastrin A [30] 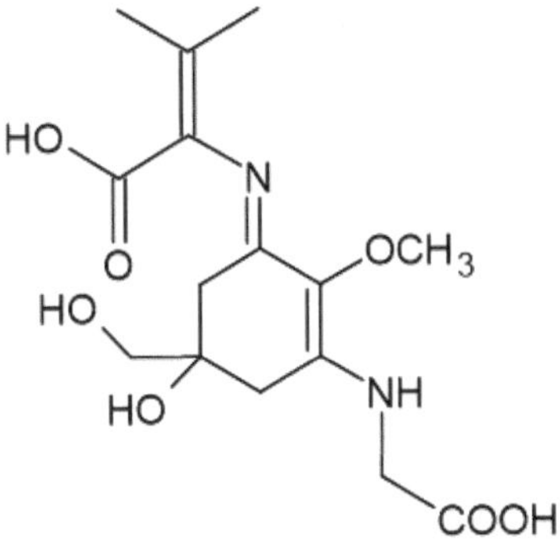	MW: 342 λ_{max} = ND ε = ND
Palythine [26] 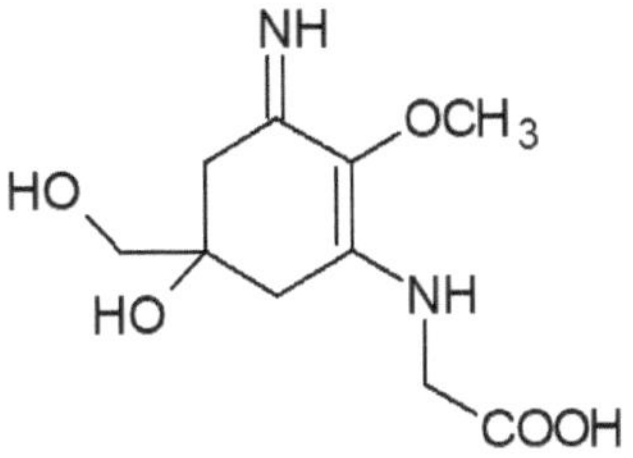	MW: 244 λ_{max} = 320 nm ε = 35,500-36,200 M^{-1} cm^{-1}

Palythine-serine [31] 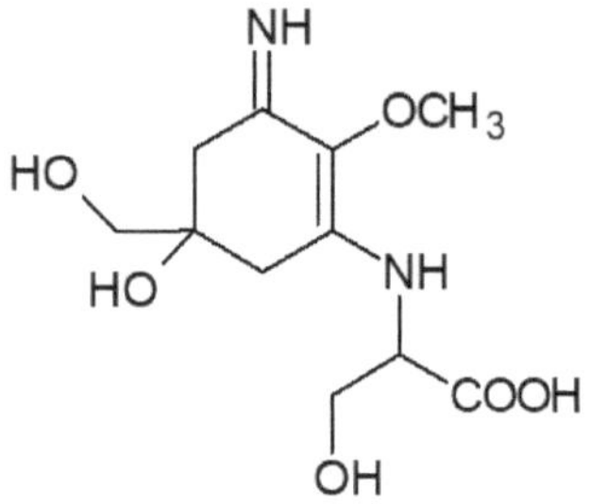	MW: 274 λ_{max} = 320 nm ε = 10,500 M^{-1} cm^{-1}
Palythine-threonine [32] 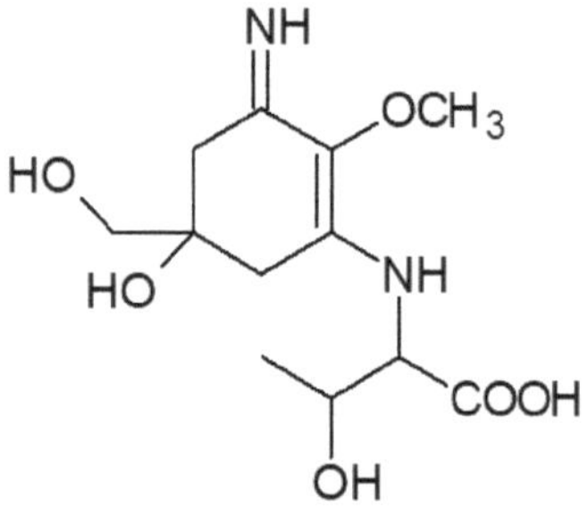	MW: 288 λ_{max} = 320 nm ε = ND
Bostrychine A [33] 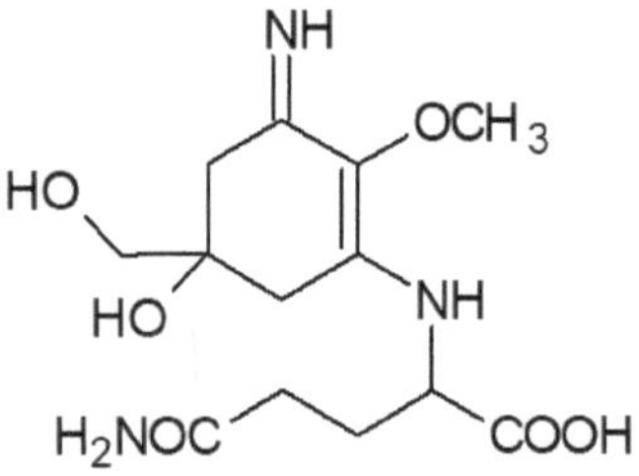	MW: 315 λ_{max} = 322 nm ε = ND
Bostrychine B [33] 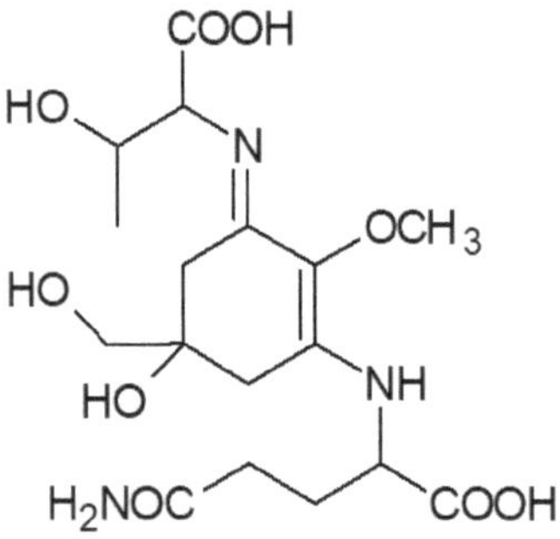	MW: 417 λ_{max} = 335 nm ε = 36,155 M^{-1} cm^{-1}

Bostrychine C [33] 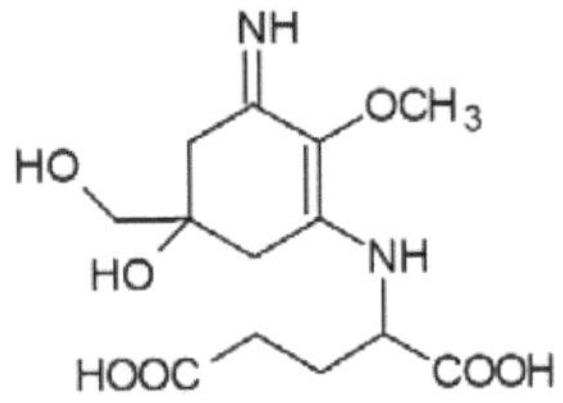	MW: 316 λ_{max} = 322 nm ε = 22,351 M^{-1} cm^{-1}
Bostrychine D [33] 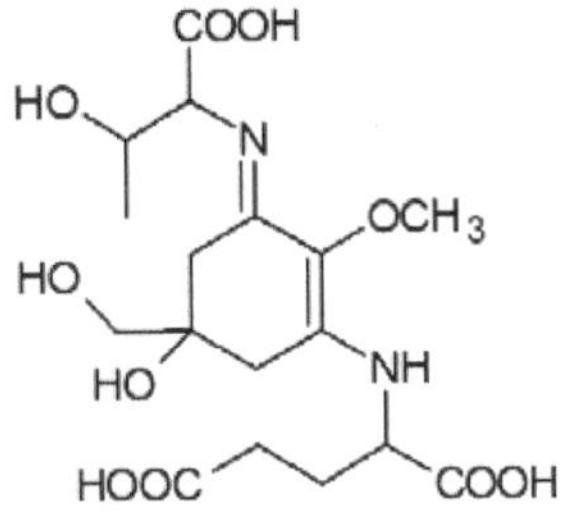	MW: 418 λ_{max} = 322 nm ε = 31,956 M^{-1} cm^{-1}
Bostrychine E [33] 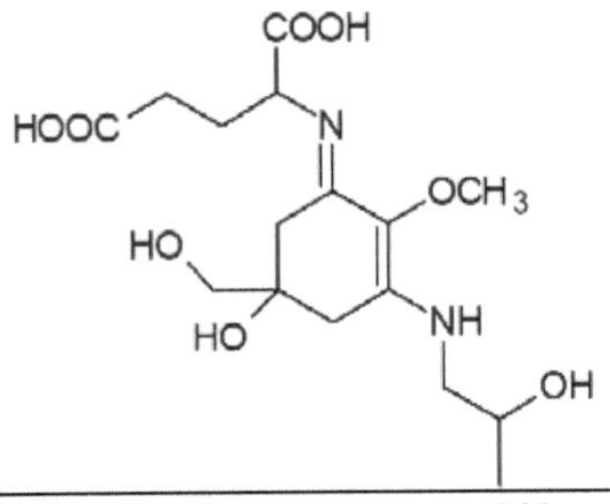	MW: 274 λ_{max} = 333 nm ε = 21,618 M^{-1} cm^{-1}
Bostrychine F [33] 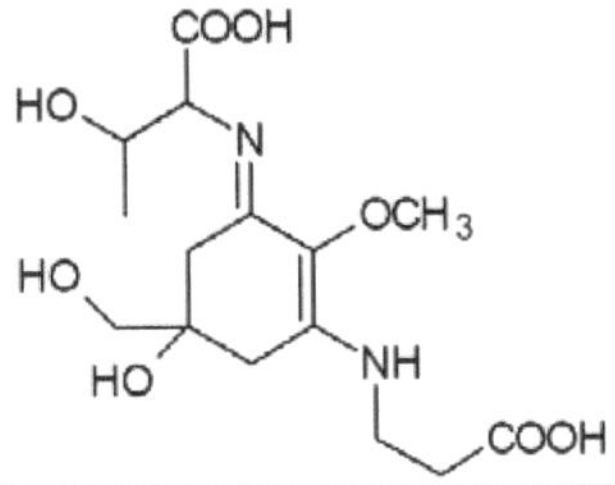	MW: 360 λ_{max} = 332 nm ε = 44,994 M^{-1} cm^{-1}

Euhalothece-362 [34] 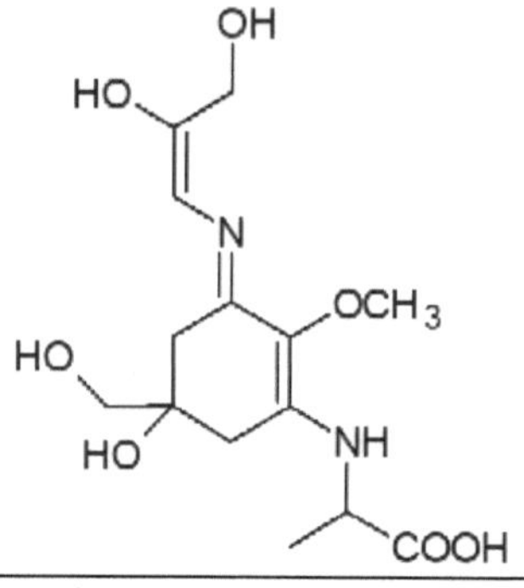	MW: 330 λ_{max} = 362 nm ε = ND
Mycosporine-methylamine-serine [32] 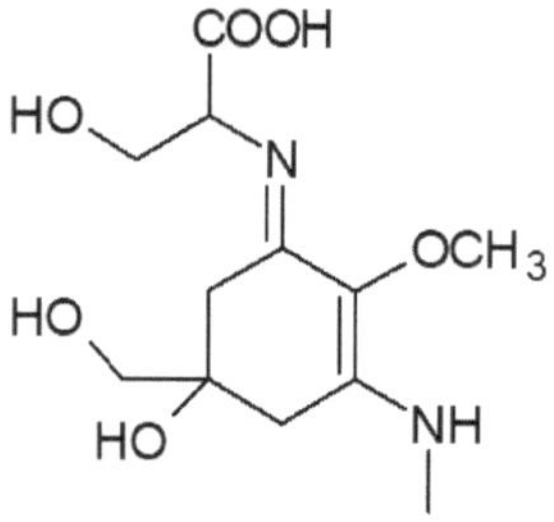	MW: 288 λ_{max} = 325 nm ε = 16,600 M^{-1} cm^{-1}
Mycosporine-methylamine-threonine [35] 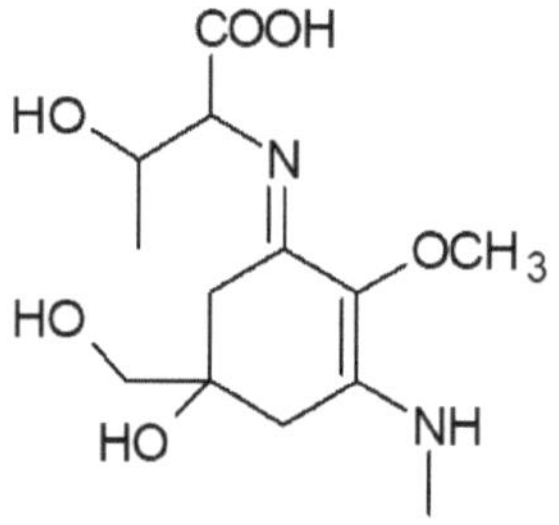	MW: 302 λ_{max} = 330 nm ε = 33,300 M^{-1} cm^{-1}
Derivatized-MAAs	
Palythine-serine sulfate [36] 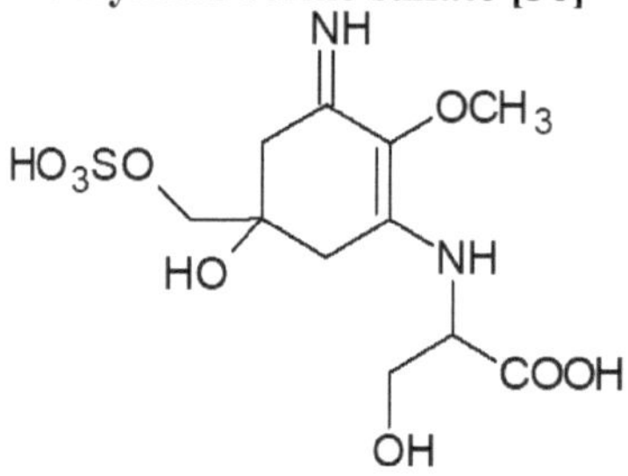	MW: 354 λ_{max} = 321 nm ε = ND

Palythine-threonine sulfate [36] 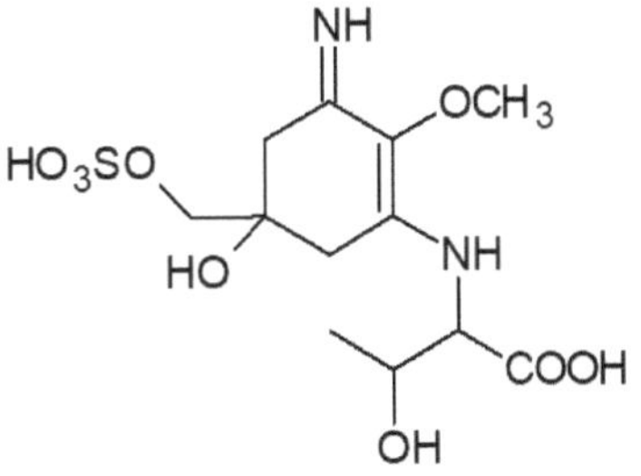	MW: 368 λ_{max} = 321 nm ε = ND
Mycosporine-glutamicol-*O*-glucoside [37] 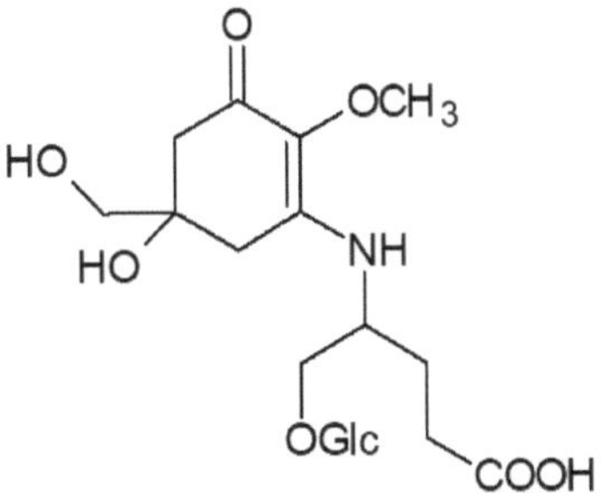	MW: 465 λ_{max} = 310 nm ε = 25,000 M^{-1} cm^{-1}
Mycosporine-glutaminol-*O*-glucoside [38] 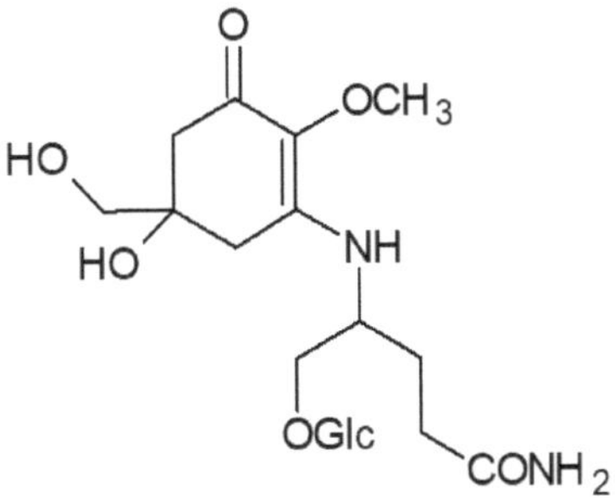	MW: 464 λ_{max} = ND ε = ND

Klebsormidin A [15] 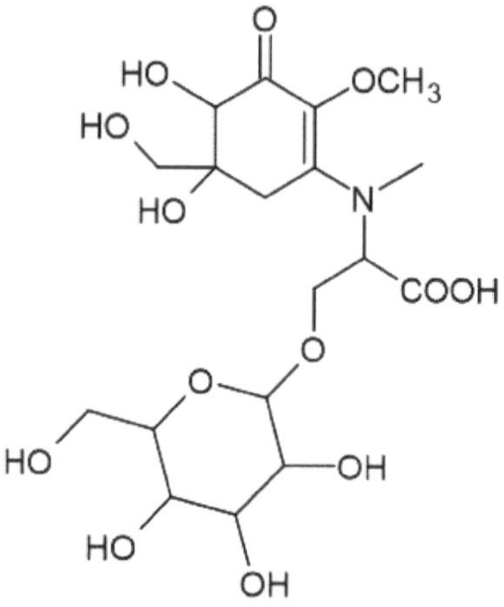	MW: 467 λ_{max} = 324 nm ε = ND
Coelastrin B [30] 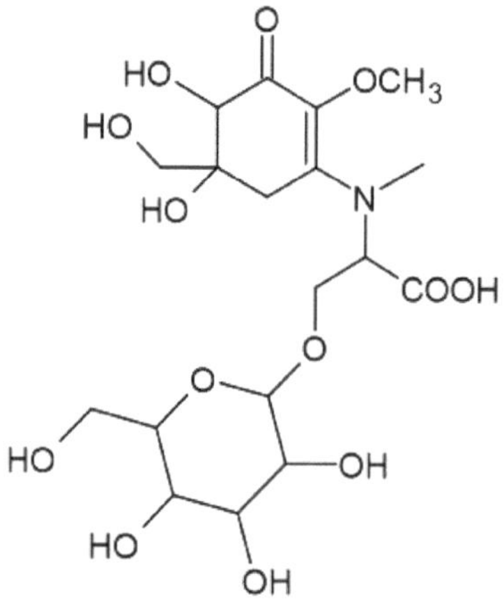	MW: 504 λ_{max} = ND ε = ND
Hexose-bound palythine-threonine [39] (450-Da MAA) 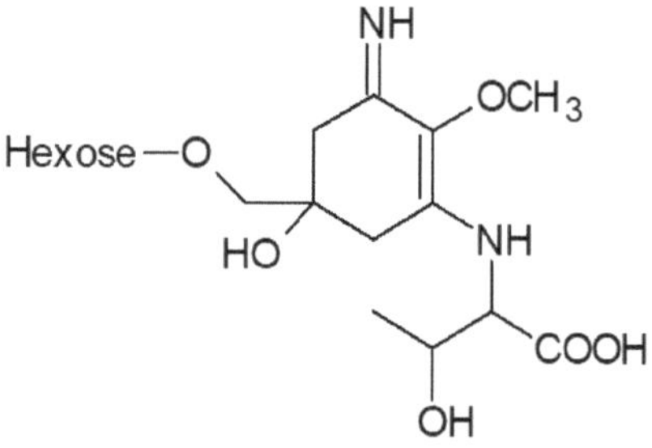	MW: 450 λ_{max} = 322 nm ε = ND

Pentose-bound shinorine (464-Da MAA) 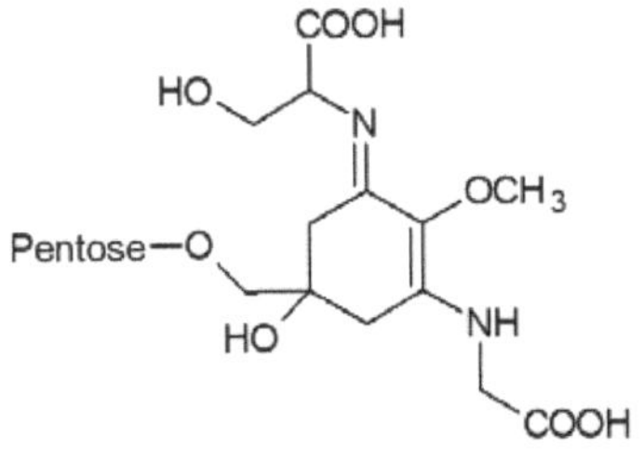	MW: 464 λ_{max} = 332 nm ε = ND
7-O-(β-arabinopyranosyl)-porphyra-334 [6] (478-Da MAA) 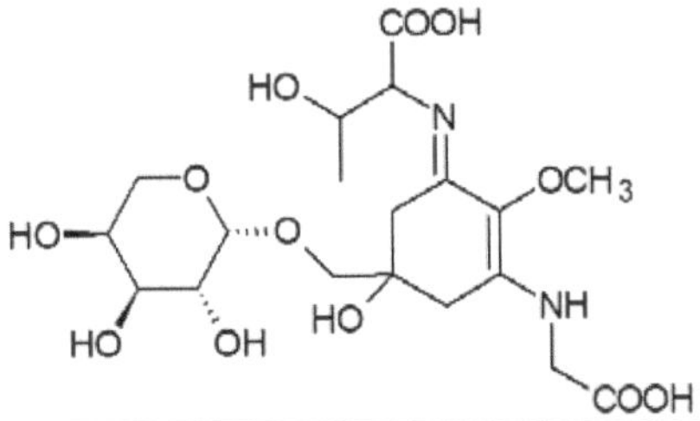	MW: 478 λ_{max} = 335 nm ε = 33,200 M^{-1} cm^{-1}
Hexose-bound porphyra-334 [39] (508-Da MAA) 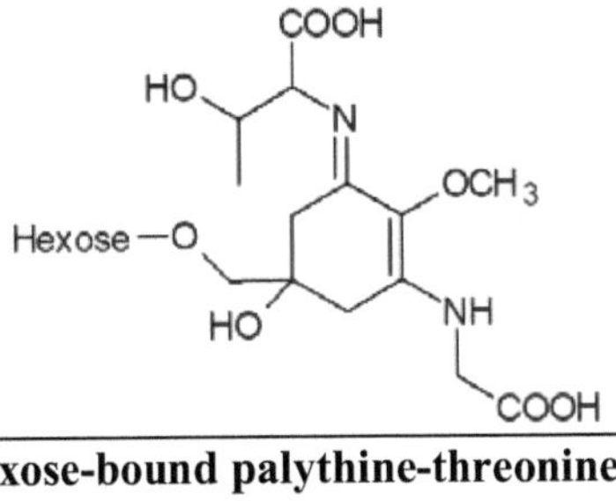	MW: 508 λ_{max} = 334 nm ε = 36,300 M^{-1} cm^{-1}
Two hexose-bound palythine-threonine [39] (612-Da MAA) 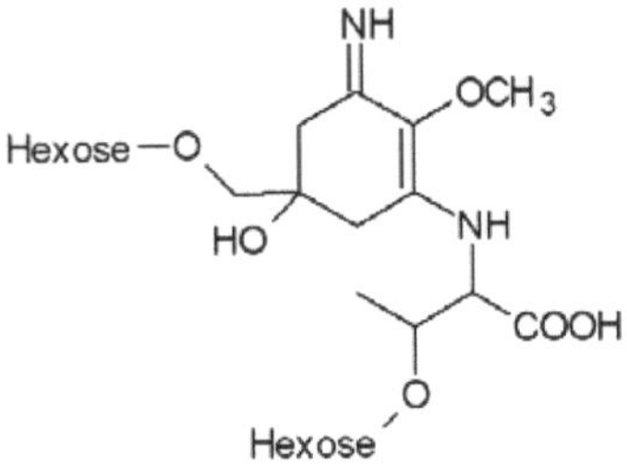	MW: 612 λ_{max} = 322 nm ε = 28,200 M^{-1} cm^{-1}

Mycosporine-4-deoxygadusolyl-ornithine [9] 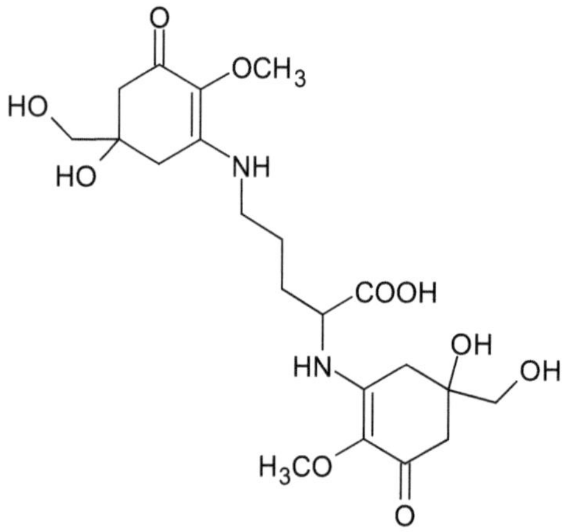	MW: 472 λ_{max} = 314 nm ε = ND
Nostoc-756 [40] (Mycosporine-2-(4-deoxygadusolyl-ornithine), 756-Da MAA) 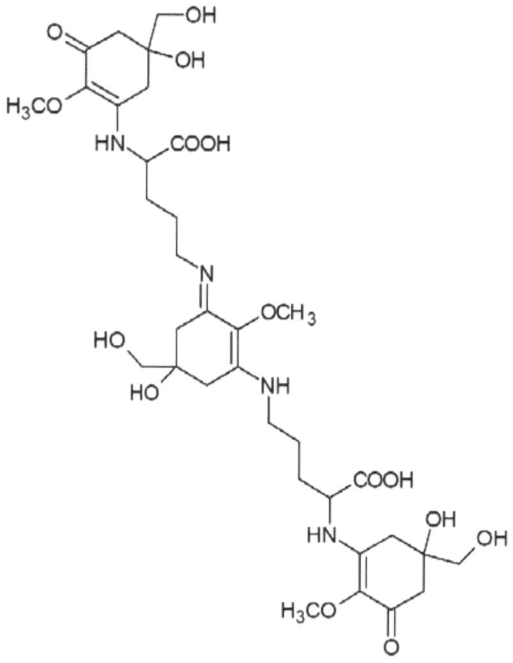	MW: 756 λ_{max} = 313 nm ε = ND

{Mycosporine-ornithine:4-deoxygadusol ornithine}-β-xylopyranosyl-β-galactopyranoside[6] (880-Da MAA) 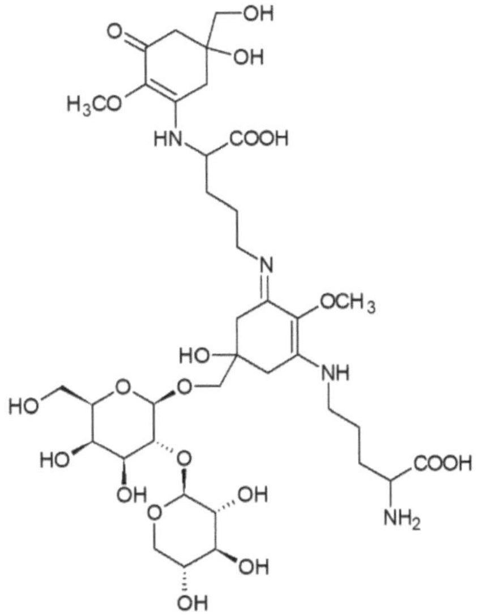	MW: 880 λ_{max} = 331 nm ε = 49,800 M^{-1} cm^{-1}
Mycosporine-2-(4-deoxygadusol-ornithine)-β-xylopyranosyl-β-galactopyranoside [6] (1050-Da MAA) 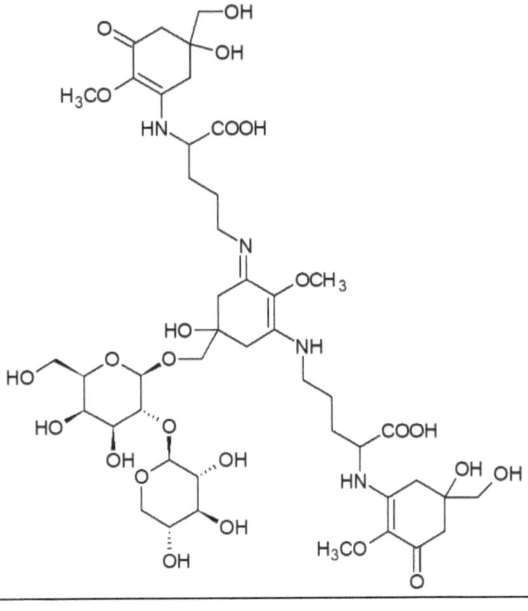	MW: 1050 λ_{max} = 312 nm ε = 58,800 M^{-1} cm^{-1}

13-*O*-(β-galactosyl)-porphyra-334 [41]	MW: 508 λ_{max} = 334 nm ε = 47,700 M^{-1} cm^{-1}

In Appendix A the list is made with reference to the information described in ref. [42]. In addition, we added information on MAAs reported after ref. [42] was published.

Although hexose-bound shinorine and hexose-bound palythine-serine have been reported in ref. [43], they were excluded from the list because the binding sites of hexose to MAAs were not clear in these compounds.

As an example of MAA structures other than those listed above, it was reported that Nagase & Co., Ltd. obtained several novel MAAs using *Streptomyces lividans* strain 1326 by introduction of MAA biosynthesis genes as an attempt for industrial synthesis of MAAs (Japanese Patent 2019-006679). These MAAs are not included in the above list due to limited structural information, but Table **1** provides information on the amino acid substituents to the parts of R_1 and R_2 of the cyclohexenimine ring for several novel MAAs and their absorption maxima.

Table 1: Substituents for novel MAAs and thier absorption maxima from JP 2019-006679.

#	Substituent R_1	Substituent R_2	Absorption Maximum (nm)
1	Alanine	Serine	331-334
2	Glycine	Allothreonine	331-334

(Table 1) cont.....

3	Alanine	Threonine	330
4	valine	Oxygen atom (Cyclohexenone ring)	310
5	Alanine	Leucine	330
6	Alanine	Isoleucine	330
7	Glycine	Leucine	330
8	Glycine	Isoleucine	330
9	Alanine	Valine	330
10	Glycine	Glutamine	336
11	Alanine	Asparagine	336
12	Glycine	Asparagine	320

REFERENCES

[1] Balskus, E.P.; Walsh, C.T. The genetic and molecular basis for sunscreen biosynthesis in cyanobacteria. *Science,* **2010**, *329*(5999), 1653-1656.
[http://dx.doi.org/10.1126/science.1193637] [PMID: 20813918]

[2] Ito, S.; Hirata, Y. Isolation and structure of a mycosporine from the zoanthid. *Tetrahedron Lett.,* **1977**, *18*(28), 2429-2430.
[http://dx.doi.org/10.1016/S0040-4039(01)83784-9]

[3] Stochaj, W.R.; Dunlap, W.C.; Shick, J.M. Two new UV-absorbing mycosporine-like amino acids from the sea anemoneAnthopleura elegantissima and the effects of zooxanthellae and spectral irradiance on chemical composition and content. *Mar. Biol.,* **1994**, *118*(1), 149-156.
[http://dx.doi.org/10.1007/BF00699229]

[4] Leite, B.; Nicholson, R.L. Mycosporine-alanine: A self-inhibitor of germination from the conidial mucilage ofColletotrichum graminicola. *Exp. Mycol.,* **1992**, *16*(1), 76-86.
[http://dx.doi.org/10.1016/0147-5975(92)90043-Q]

[5] Saha, S.; Sen, A.; Mandal, S.; Adhikary, S.P.; Rath, J. Mycosporine-alanine, an oxo-mycosporine, protect Hassallia byssoidea from high UV and solar irradiation on the stone monument of Konark. *J. Photochem. Photobiol. B,* **2021**, *224*112302.
[http://dx.doi.org/10.1016/j.jphotobiol.2021.112302] [PMID: 34537544]

[6] Nazifi, E.; Wada, N.; Asano, T.; Nishiuchi, T.; Iwamuro, Y.; Chinaka, S.; Matsugo, S.; Sakamoto, T. Characterization of the chemical diversity of glycosylated mycosporine-like amino acids in the terrestrial cyanobacterium Nostoc commune. *J. Photochem. Photobiol. B,* **2015**, *142*, 154-168. [http://dx.doi.org/10.1016/j.jphotobiol.2014.12.008] [PMID: 25543549]

[7] Arpin, N.; Curt, R.; Fravre-Bonvin, J. Mycosporines: Review and new data concerning their structure. *Revue de Mycologie,* **1979**, *43*, 247-257.

[8] Katoch, M.; Mazmouz, R.; Chau, R.; Pearson, L.A.; Pickford, R.; Neilan, B.A. Heterologous production of cyanobacterial mycosporine-like amino acids mycosporine-ornithine and mycosporine-lysine in Escherichia coli. *Appl. Environ. Microbiol.,* **2016**, *82*(20), 6167-6173. [http://dx.doi.org/10.1128/AEM.01632-16] [PMID: 27520810]

[9] Zhang, Z.C.; Wang, K.; Hao, F.H.; Shang, J.L.; Tang, H.R.; Qiu, B.S. New types of ATP-grasp ligase are associated with the novel pathway for complicated mycosporine-like amino acid production in desiccation□tolerant cyanobacteria. *Environ. Microbiol.,* **2021**, *23*(11), 6420-6432. [http://dx.doi.org/10.1111/1462-2920.15732] [PMID: 34459073]

[10] White, J.D.; Cammack, J.H.; Sakuma, K.; Rewcastle, G.W.; Widener, R.K. Transformations of quinic acid. Asymmetric synthesis and absolute configuration of mycosporin I and mycosporin-gly. *J. Org. Chem.,* **1995**, *60*(12), 3600-3611. [http://dx.doi.org/10.1021/jo00117a008]

[11] Young, H.; Patterson, V.J. A uv protective compound from Glomerella cingulata—a mycosporine. *Phytochemistry,* **1982**, *21*(5), 1075-1077. [http://dx.doi.org/10.1016/S0031-9422(00)82419-X]

[12] Lunel, M-C.; Arpin, N.; Favre-Bonvin, J. Structure of normycosporin glutamine, a new compound isolated from Pyronema omphalodes (Bull ex Fr.) fuckel. *Tetrahedron Lett.,* **1980**, *21*, 4715-4716. [http://dx.doi.org/10.1016/0040-4039(80)88101-9]

[13] Bernillon, J.; Parussini, E.; Letoublon, R.; Favre-Bonvin, J.; Arpin, N. Flavin-mediated photolysis of mycosporines. *Phytochemistry,* **1990**, *29*(1), 81-84. [http://dx.doi.org/10.1016/0031-9422(90)89015-2]

[14] Pittet, J.L.; Bouillant, M.L.; Bernillon, J.; Arpin, N.; Favre-Bonvin, J. The presence of reduced-glutamine mycosporines, new molecules, in several Deuteromycetes. *Tetrahedron Lett.,* **1983**, *24*, 65-68. [http://dx.doi.org/10.1016/S0040-4039(00)81328-3]

[15] Hartmann, A.; Glaser, K.; Holzinger, A.; Ganzera, M.; Karsten, U. Klebsormidin A and B, two new UV-sunscreen compounds in green microalgal Interfilum and Klebsormidium Species (Streptophyta) from terrestrial habitats. *Front. Microbiol.,* **2020**, *11*, 499. [http://dx.doi.org/10.3389/fmicb.2020.00499] [PMID: 32292396]

[16] Tsujino, I.; Yabe, K.; Sekikawa, I. Isolation and structure of a new amino acid, shinorine, from the red alga Chondrus yendoi. *Bot. Mar.,* **1980**, *23*(1), 65-67. [http://dx.doi.org/10.1515/botm.1980.23.1.63]

[17] Takano, S.; Nakanishi, A.; Uemura, D.; Hirata, Y. Isolation and structure of a 334 nm UV-absorbing substance, porphyra-334 from the red alga Porphyra tenera Kjellman. *Chem. Lett.,* **1979**, *8*(419-420).

[18] Kedar, L.; Kashman, Y.; Oren, A. Mycosporine-2-glycine is the major mycosporine-like amino acid in a unicellular cyanobacterium (*Euhalothece* sp.) isolated from a gypsum crust in a hypersaline saltern pond. *FEMS Microbiol. Lett.,* **2002**, *208*(2), 233-237. [http://dx.doi.org/10.1111/j.1574-6968.2002.tb11087.x] [PMID: 11959442]

[19] Nakamura, H.; Kobayashi, J.; Hirata, Y. Isolation and structure of a 330 nm UV-absorbing substance, asterina-330 from the starfish Asterina pectinifera. *Chem. Lett.,* **1981**, *10*(10), 1413-1414. [http://dx.doi.org/10.1246/cl.1981.1413]

[20] Bandaranayake, W.M.; Bemis, J.E.; Bourne, D.J. Ultraviolet absorbing pigments from the marine sponge Dysidea herbacea: Isolation and structure of a new mycosporine. *Comp. Biochem. Physiol.,* **1996**, *115C*(3), 281-286.

[21] Grant, P.T.; Middleton, C.; Plack, P.A.; Thomson, R.H. The isolation of four aminocyclohexenimines (mycosporines) and a structurally related derivative of cyclohexane-1:3-dione (gadusol) from the brine shrimp, Artemia. *Comp. Biochem. Physiol. B,* **1985**, *80*(4), 755-759. [http://dx.doi.org/10.1016/0305-0491(85)90457-2]

[22] Karentz, D.; McEuen, F.S.; Land, M.C.; Dunlap, W.C. Survey of mycosporine-like amino acid compounds in Antarctic marine organisms: Potential protection from ultraviolet exposure. *Mar. Biol.,* **1991**, *108*(1), 157-166. [http://dx.doi.org/10.1007/BF01313484]

[23] Miyamoto, K.T.; Komatsu, M.; Ikeda, H. Discovery of gene cluster for mycosporine-like amino acid biosynthesis from Actinomycetales microorganisms and production of a novel mycosporine-like amino acid by heterologous expression. *Appl. Environ. Microbiol.,* **2014**, *80*(16), 5028-5036. [http://dx.doi.org/10.1128/AEM.00727-14] [PMID: 24907338]

[24] Chen, M.; Rubin, G.M.; Jiang, G.; Raad, Z.; Ding, Y. Biosynthesis and heterologous production of mycosporine-like amino acid palythines. *J. Org. Chem.,* **2021**, *86*(16), 11160-11168. [http://dx.doi.org/10.1021/acs.joc.1c00368] [PMID: 34006097]

[25] Sekikawa, I.; Kubota, C.; Hiraoki, T.; Tsujino, I. Isolation and structure of a 357 nm UV-absorbing substance, usujirene, from the red alga Palmaria palmata (L.) O. Kuntze. *Jpn. J. Phycol.,* **1986**, *34*, 185-188.

[26] Takano, S.; Uemura, D.; Hirata, Y. Isolation and structure of two new amino acids, palythinol and palythene, from the zoanthid. *Tetrahedron Lett.,* **1978**, *19*(49), 4909-4912. [http://dx.doi.org/10.1016/S0040-4039(01)85768-3]

[27] Kobayashi, J.; Nakamura, H.; Hirata, Y. Isolation and structure of a UV-absorbing substance 337 from the ascidian Halocynthia roretzi. *Tetrahedron Lett.,* **1981**, *22*, 3001-3002. [http://dx.doi.org/10.1016/S0040-4039(01)81811-6]

[28] Kamio, M.; Kicklighter, C.E.; Nguyen, L.; Germann, M.W.; Derby, C.D. Isolation and structural elucidation of novel mycosporine-like amino acids as alarm cues in the defensive ink secretion of the sea hare Aplysia colifornica. *Helv. Chim. Acta,* **2011**, *94*(6), 1012-1018. [http://dx.doi.org/10.1002/hlca.201100117]

[29] Werner, N.; Orfanoudaki, M.; Hartmann, A.; Ganzera, M.; Sommaruga, R. Low temporal dynamics of mycosporine-like amino acids in benthic cyanobacteria from an alpine lake. *Freshw. Biol.,* **2021**, *66*(1), 169-176.

[http://dx.doi.org/10.1111/fwb.13627] [PMID: 33510548]

[30] Zaytseva, A.; Chekanov, K.; Zaytsev, P.; Bakhareva, D.; Gorelova, O.; Kochkin, D.; Lobakova, E. Sunscreen effect exerted by secondary carotenoids and mycosporine-like amino acids in the aeroterrestrial Chlorophyte Coelastrella rubescens under high light and UV-A irradiation. *Plants,* **2021**, *10*(12), 2601. [http://dx.doi.org/10.3390/plants10122601] [PMID: 34961072]

[31] Teai, T.T.; Raharivelomanana, P.; Bianchini, J.P.; Faure, R.; Martin, P.M.V.; Cambon, A. Structure de deux nouvelles iminomycosporines isolées de Pocillopora eydouxi. *Tetrahedron Lett.,* **1997**, *38*(33), 5799-5800. [http://dx.doi.org/10.1016/S0040-4039(97)01281-1]

[32] Carignan, M.O.; Cardozo, K.H.M.; Oliveira-Silva, D.; Colepicolo, P.; Carreto, J.I. Palythine–threonine, a major novel mycosporine-like amino acid (MAA) isolated from the hermatypic coral Pocillopora capitata. *J. Photochem. Photobiol. B,* **2009**, *94*(3), 191-200. [http://dx.doi.org/10.1016/j.jphotobiol.2008.12.001] [PMID: 19128981]

[33] Orfanoudaki, M.; Hartmann, A.; Alilou, M.; Gelbrich, T.; Planchenault, P.; Derbré, S.; Schinkovitz, A.; Richomme, P.; Hensel, A.; Ganzera, M. Absolute configuration of mycosporine-like amino acids, their wound healing properties and *in vitro* anti-aging effects. *Mar. Drugs,* **2019**, *18*(1), 35. [http://dx.doi.org/10.3390/md18010035] [PMID: 31906052]

[34] Volkmann, M.; Gorbushina, A.A.; Kedar, L.; Oren, A. Structure of euhalothece-362, a novel red-shifted mycosporine-like amino acid, from a halophilic cyanobacterium (Euhalothece sp.). *FEMS Microbiol. Lett.*, **2006**, *258*(1), 50-54. [http://dx.doi.org/10.1111/j.1574-6968.2006.00203.x] [PMID: 16630254]

[35] Wu Won, J.; Rideout, J.A.; Chalker, B.E. Isolation and structure of a novel mycosporine-like amino acid from the reef-building corals Pocillopora damicornis and Stylophora pistillata. *Tetrahedron Lett.*, **1995**, *36*(29), 5255-5256. [http://dx.doi.org/10.1016/00404-0399(50)0950H-]

[36] Wu Won, J.J.; Chalker, B.E.; Rideout, J.A. Two new UV-absorbing compounds from Stylophora pistillata: Sulfate esters of mycosporine-like amino acids. *Tetrahedron Lett.*, **1997**, *38*(14), 2525-2526. [http://dx.doi.org/10.1016/S0040-4039(97)00391-2]

[37] Bouillant, M.L.; Pittet, J.L.; Bernillon, J.; Favre-Bonvin, J.; Arpin, N. Mycosporins from Ascochyta pisi, Cladosporium herbarum and Septoria nodorum. *Phytochemistry,* **1981**, *20*(12), 2705-2707. [http://dx.doi.org/10.1016/0031-9422(81)85272-7]

[38] Bernillon, J.; Bouillant, M.L.; Pittet, J.L.; Favre-Bonvin, J.; Arpin, N. Mycosporine glutamine and related mycosporines in the fungus Pyronema omphalodes. *Phytochemistry,* **1984**, *23*(5), 1083-1087. [http://dx.doi.org/10.1016/S0031-9422(00)82614-X]

[39] Nazifi, E.; Wada, N.; Yamaba, M.; Asano, T.; Nishiuchi, T.; Matsugo, S.; Sakamoto, T. Glycosylated porphyra-334 and palythine-threonine from the terrestrial cyanobacterium Nostoc commune. *Mar. Drugs,* **2013**, *11*(9), 3124-3154. [http://dx.doi.org/10.3390/md11093124] [PMID: 24065157]

[40] Sakamoto, T.; Hashimoto, A.; Yamaba, M.; Wada, N.; Yoshida, T.; Inoue-Sakamoto, K.; Nishiuchi, T.; Matsugo, S. Four chemotypes of the terrestrial cyanobacterium *Nostoc commune* characterized by differences in the mycosporine□like amino acids. *Phycol. Res.,* **2019**, *67*(1), 3-11. [http://dx.doi.org/10.1111/pre.12333]

[41] Ishihara, K.; Watanabe, R.; Uchida, H.; Suzuki, T.; Yamashita, M.; Takenaka, H.; Nazifi, E.; Matsugo, S.; Yamaba, M.; Sakamoto, T. Novel glycosylated mycosporine-like amino acid, 13- O -(- -galactosyl)-porphyra-334, from the edible cyanobacterium Nostoc sphaericum -protective activity on human keratinocytes from UV light. *J. Photochem. Photobiol. B,* **2017**, *172*, 102-108. [http://dx.doi.org/10.1016/j.jphotobiol.2017.05.019] [PMID: 28544967]

[42] Wada, N.; Sakamoto, T.; Matsugo, S. Mycosporine-like amino acids and their derivatives as natural antioxidants. *Antioxidants,* **2015**, *4*(3), 603-646. [http://dx.doi.org/10.3390/antiox4030603] [PMID: 26783847]

[43] D'Agostino, P.M.; Javalkote, V.S.; Mazmouz, R.; Pickford, R.; Puranik, P.R.; Neilan, B.A. Comparative profiling and discovery of novel glycosylated mycosporine-like amino acids in two strains of the cyanobacterium Scytonema cf. crispum. *Appl. Environ. Microbiol.,* **2016**, *82*(19), 5951-5959. [http://dx.doi.org/10.1128/AEM.01633-16] [PMID: 27474710]

Appendix B

Classification and Molecular Structures of Amino Acids

1. Proteinogenic Amino Acids

Classification		Amino acid	Abbr. 3-ltrs	Abbr. 1-ltr	Structure	MW (MF)
Neutral amino acids	Aliphatic amino acids	Glycine	Gly	G		75 ($C_2H_5NO_2$)
		Alanine	Ala	A		89 ($C_3H_7NO_2$)
		Valine	Val	V		117 ($C_5H_{11}NO_2$)
		Leucine	Leu	L		131 ($C_6H_{13}NO_2$)
		Isoleucine	Ile	I		131 ($C_2H_5NO_2$)
	Oxyamino acids	Serine	Ser	S		105 ($C_3H_7NO_3$)
		Threonine	Thr	T		119 ($C_4H_9NO_3$)
	Amino acids containing sulfur	Cysteine	Cys	C		121 ($C_3H_7NO_2S$)
		Cystine	$(Cys)_2$	—		240 ($C_6H_{12}N_2O_4S_2$)

Hakuto Kageyama

		Methionine	Met	M	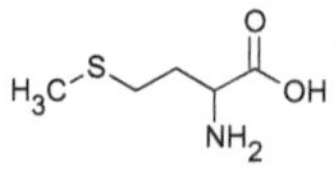	149 ($C_5H_{11}NO_2S$)
	Aromatic amino acids	Phenylalanine	Phe	F	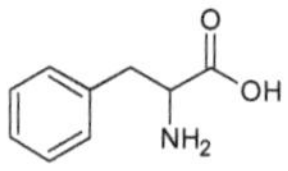	165 ($C_9H_{11}NO_2$)
		Tyrosine	Tyr	Y	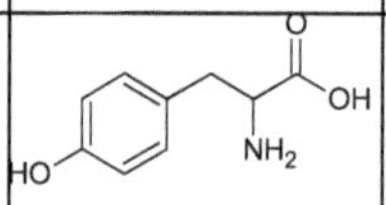	181 ($C_9H_{11}NO_3$)
		Tryptophan	Trp	W	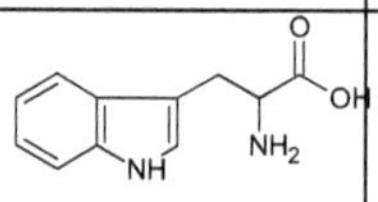	204 ($C_{11}H_{12}N_2O_2$)
	Imino acid	Proline	Pro	P	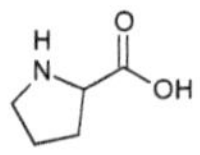	115 ($C_5H_9NO_2$)
		Hydroxyproline	Hyp	–	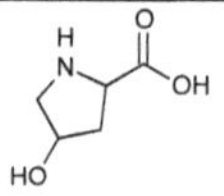	131 ($C_5H_9NO_3$)
	Acid amide	Asparagine	Asn	N	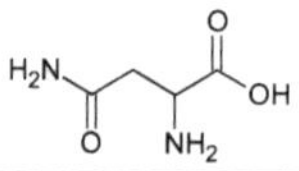	132 ($C_4H_8N_2O_3$)
		Glutamine	Gln	Q	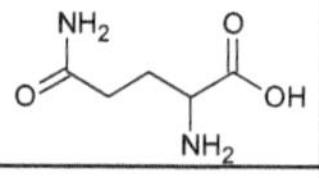	146 ($C_5H_{10}N_2O_3$)
Acidic amino acids		Aspartic acid	Asp	D	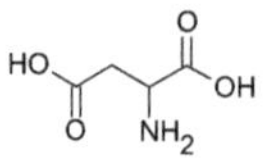	133 ($C_4H_7NO_4$)
		Glutamic Acid	Glu	E	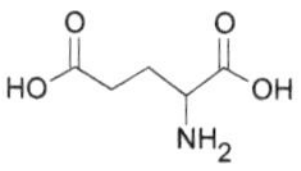	147 ($C_5H_9NO_4$)
Basic amino acids		Lysine	Lys	K	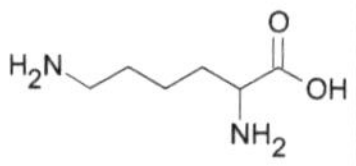	146 ($C_6H_{14}N_2O_2$)
		Histidine	His	H	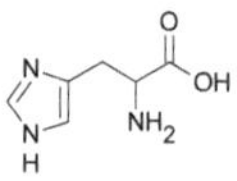	155 ($C_6H_9N_3O_2$)

	Arginine	Arg	R	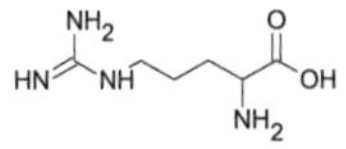	174 ($C_6H_{14}N_4O_2$)

2. Representitive Amino Acids other than Protein Constituent

Classification		Amino acid	Abbr.	Structure	MW (MF)
α-Amino acids	Aliphatic amino acids	α-Aminobutyric acid	Abu	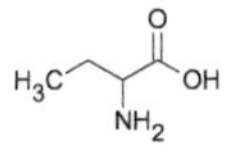	103 ($C_4H_9NO_2$)
		Norvaline	Nva	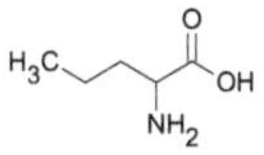	117 ($C_5H_{11}NO_2$)
		Norleucine	Nle	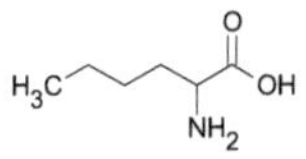	131 ($C_6H_{13}NO_2$)
		Homoleucine	Hle	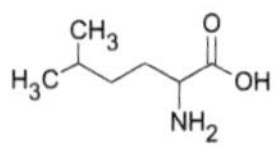	145 ($C_7H_{15}NO_2$)
	Oxyamino acids	Homoserine	Hse	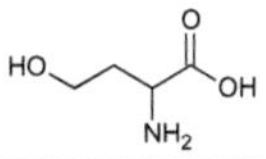	119 ($C_4H_9NO_3$)
	Amino acids containing sulfur	Homocysteine	Hcy	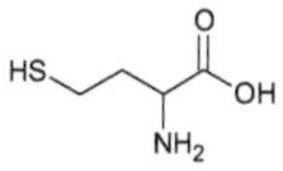	135 ($C_4H_9NO_2S$)
		Cysteic acid	—	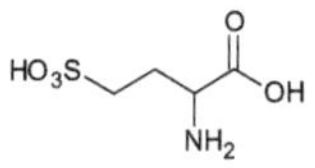	169 ($C_3H_7NO_5S$)
	Aromatic amino acids	3,4-Dihydroxyphenylalanine	DOPA	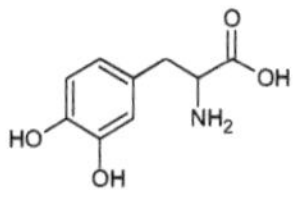	197 ($C_9H_{11}NO_4$)
	Basic amino acids	Ornithine	Orn	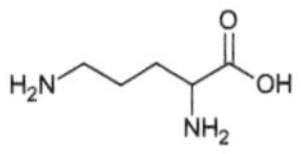	132 ($C_5H_{11}N_2O_2$)
β-Amino acid		β-Alanine	β-Ala	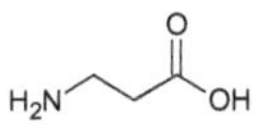	89 ($C_3H_7NO_2$)

γ-Amino acid	γ-Aminobutyric acid	GABA	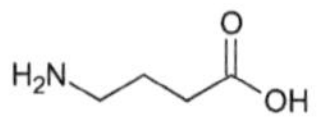 H_2N, O, OH	103 ($C_4H_9NO_2$)
Amino sulfonic acid	Taurine	Tau	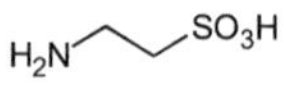 H_2N, SO_3H	125 ($C_2H_7NO_3S$)

Appendix C

Correlation Diagram of the Molecular Structure of MAAs

1. Overall view

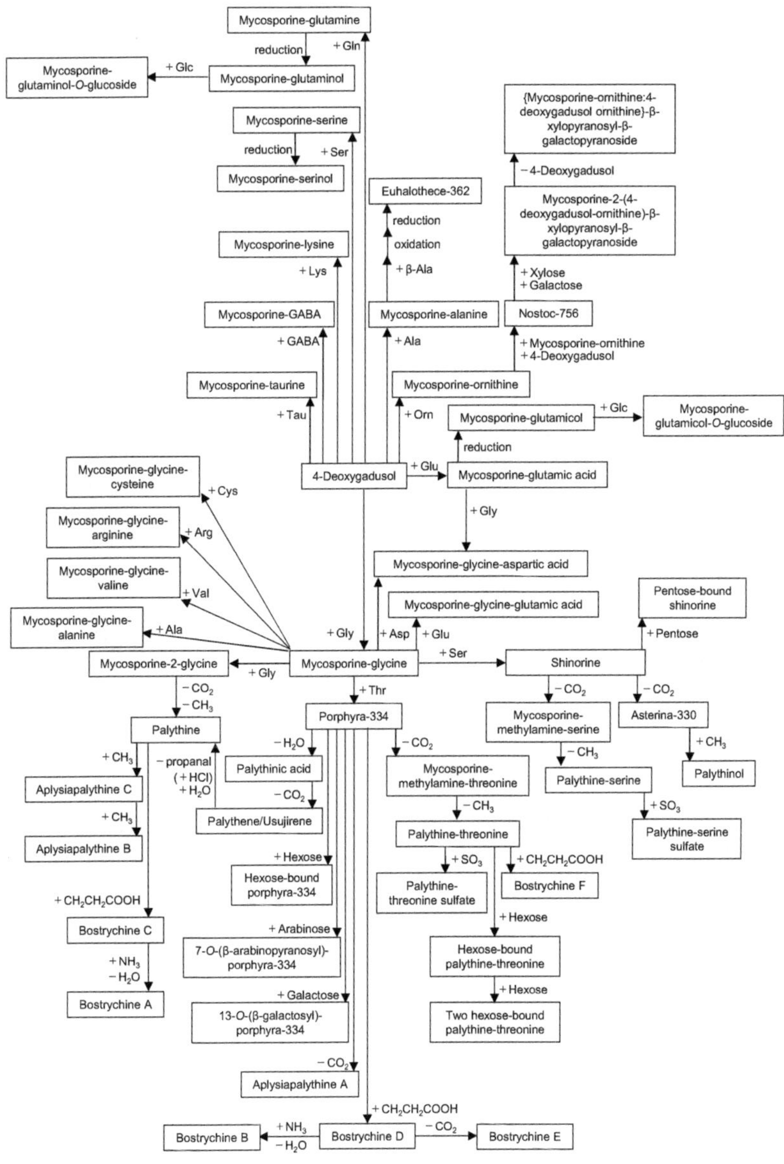

Hakuto Kageyama

2. Correlation diagram centered on 4-deoxygadusol (bottom of the overall diagram)

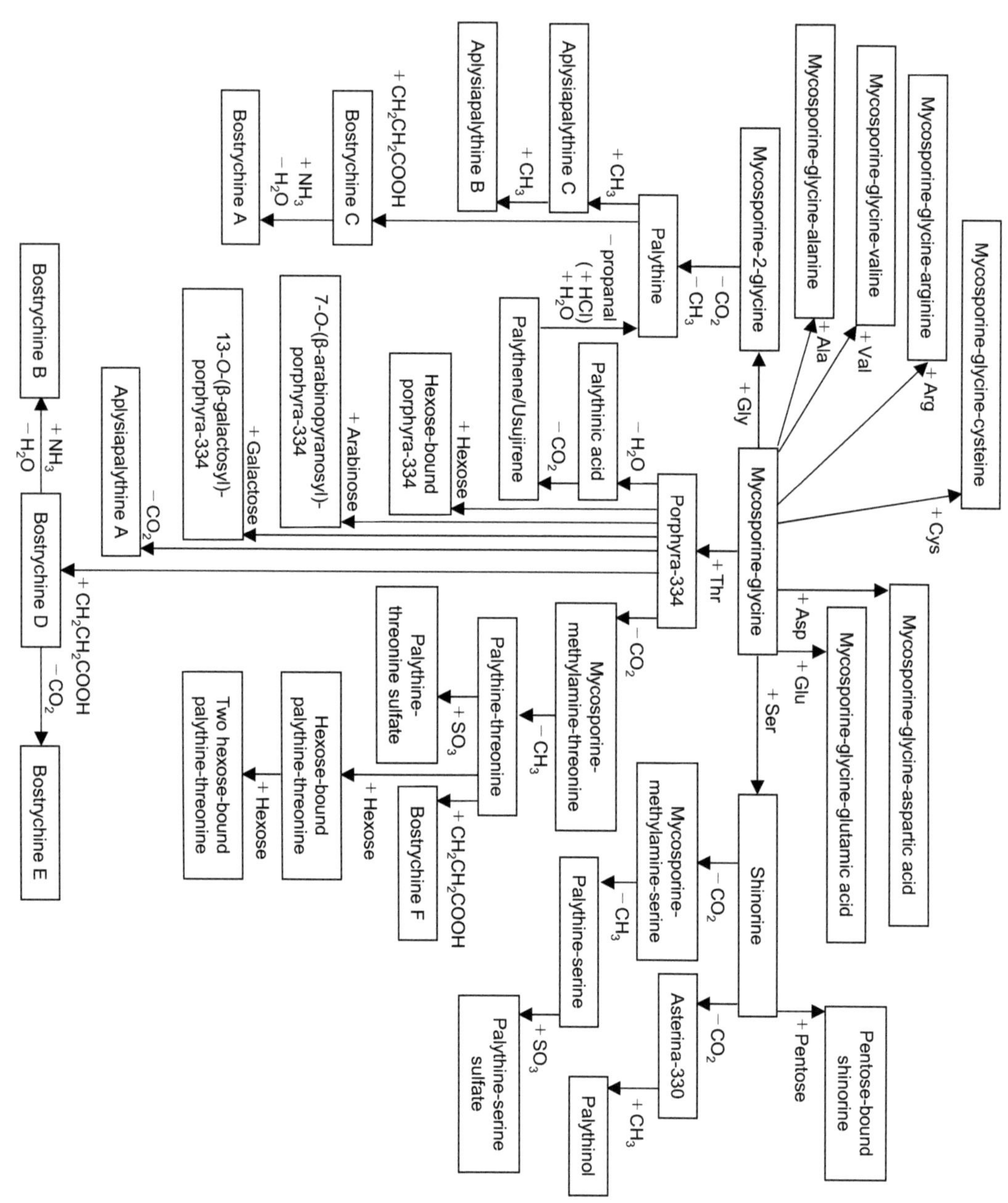

3. Correlation diagram centered on mycosporine-glycine (upper part of the overall diagram)

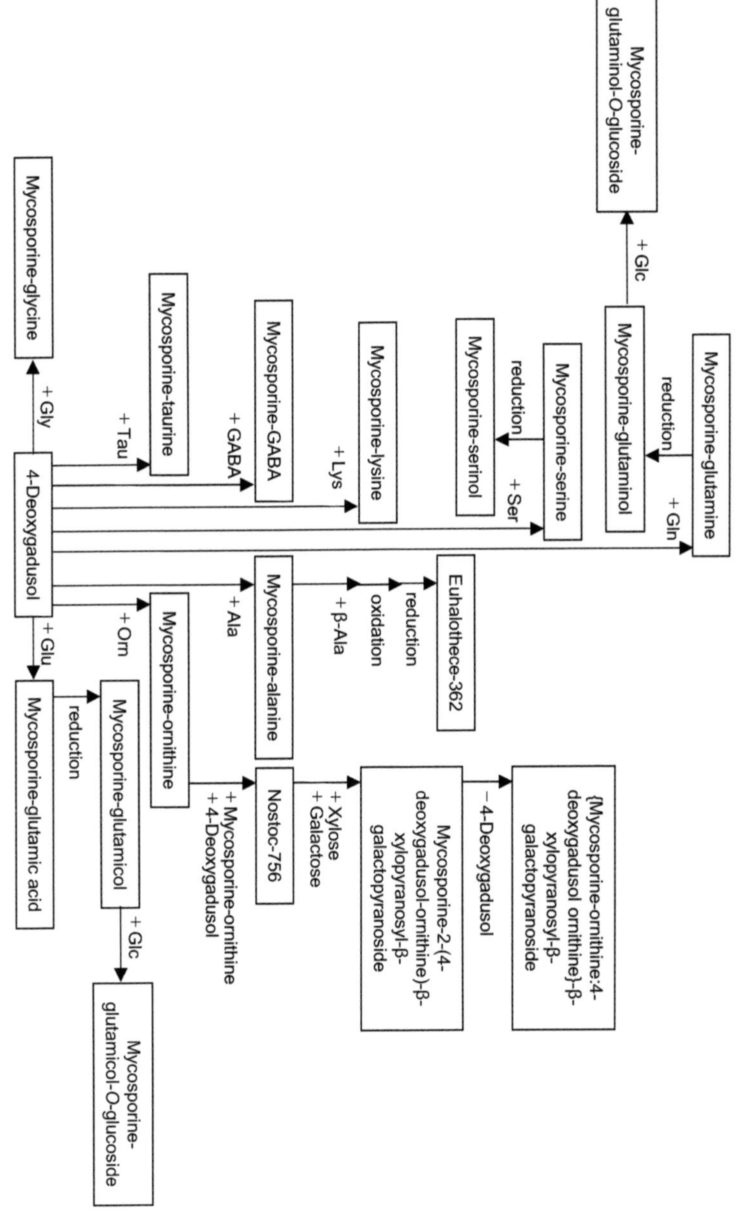

An Introduction to Mycosporine-Like Amino Acids, 2023, 144-144

Appendix D

Position Numbering of Carbon Atoms in the Molecular Structures of MAAs

1. Cyclohexenone structure

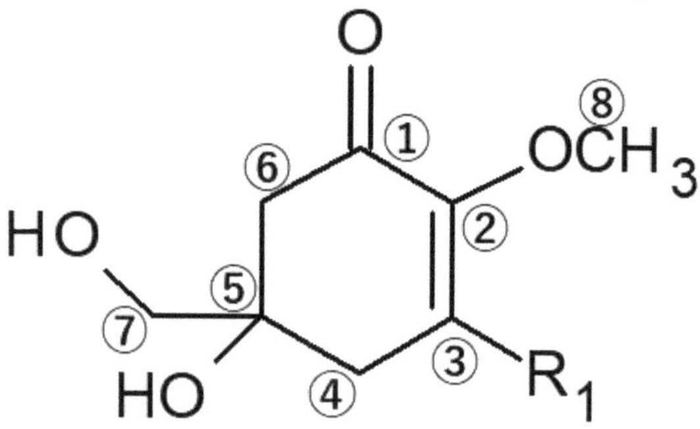

2. Cyclohexenimine structure

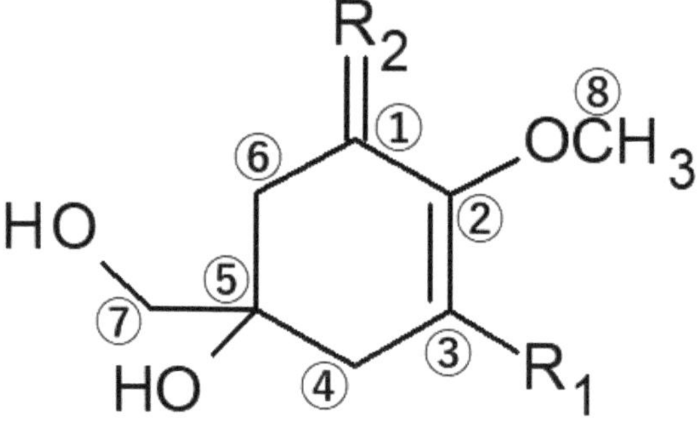

Hakuto Kageyama

Appendix E

Outline of the Scytonemin Biosynthetic Pathway

The genes involved in scytonemin biosynthesis were first reported in 2007 [1]. A cluster of 18 genes (*NpR1276–NpR1259*) involved in the scytonemin biosynthetic pathway was identified by random transposon mutagenesis in the cyanobacterium *Nostoc punctiforme* ATCC 29133. This gene cluster was well conserved among various cyanobacterial strains [2]. Scytonemin is thought to be synthesized from derivatives of the aromatic amino acids tryptophan and tyrosine. In fact, the gene cluster identified by *N. punctiforme* included 8 genes associated with the aromatic amino acid biosynthesis pathway (the tyrosine biosynthesis gene *NpR1269* as *tyrA*; tryptophan biosynthesis genes *NpR1266*, *NpR1265*, *NpR1264*, *NpR1262*, and *NpR1261* as *trpE*, *trpC*, *trpA*, *trpB*, and *trpD*, respectively; and shikimate pathway-related genes *NpR1267* and *NpR1260* as *aroB* and *aroG*, respectively).

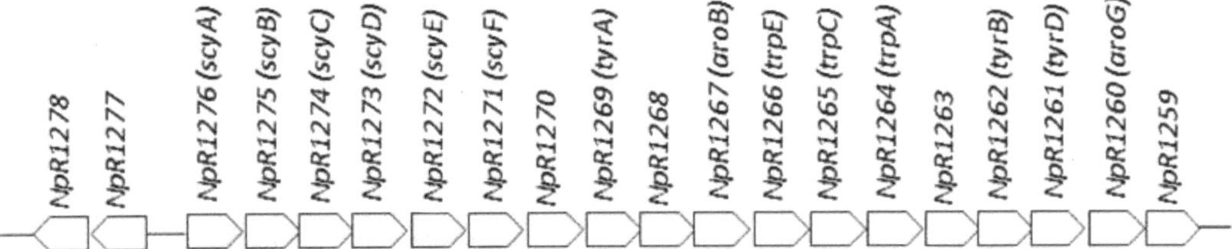

Fig. (1). Genes involved in scytonemin biosynthetic pathway.

In the scytonemin synthesis gene cluster of *N. punctiforme*, the six genes *NpR1276–NpR1271* are thought to catalyze the core reactions of scytonemin biosynthesis. They were named *scyA–scyF*, respectively. ScyB, which is similar to NADH-dependent oxidoreductase, was identified as an enzyme responsible for the early reaction stages of scytonemin biosynthesis. It is thought to promote oxidative deamination of tryptophan and to synthesize indole-3-pyruvic acid (I3P). I3P is one of the precursors required for the monomeric polycyclic alkaloids of scytonemin. The second precursor compound, *p*-hydroxyphenylpyruvic acid (HPP), is believed to be converted from prephenic acid by TyrA encoded by *NpR1269*. ScyA, which is similar to acetolactic acid synthase, promotes the condensation reaction between I3P and HPP, as revealed by *in vitro* experiments [3]. The product of the condensation reaction is cyclized and decarboxylated by ScyC [4]. The monomer compound thus obtained is thought to be dimerized by ScyD, ScyE, and ScyF,

Hakuto Kageyama

but the detailed reaction mechanism of these proteins is still unknown. Given that signal domains are present in the amino acid sequences of ScyD, ScyE, and ScyF, these proteins are thought to be present in the periplasmic space. Therefore, scytonemin biosynthesis is considered to be compartmentalized in cyanobacterial cells. That is, the synthesis of the monomer precursor compound in the early stage and the dimerization reaction in the late stage may occur in the cytoplasmic space and the periplasmic space, respectively. In addition, genes involved in the two-component signal transduction pathway (*NpR1277*/*NpR1278*) located directly upstream of the scytonemin biosynthesis gene cluster were identified, and NpR1278 was essential for scytonemin biosynthesis in *N. punctiforme* [5].

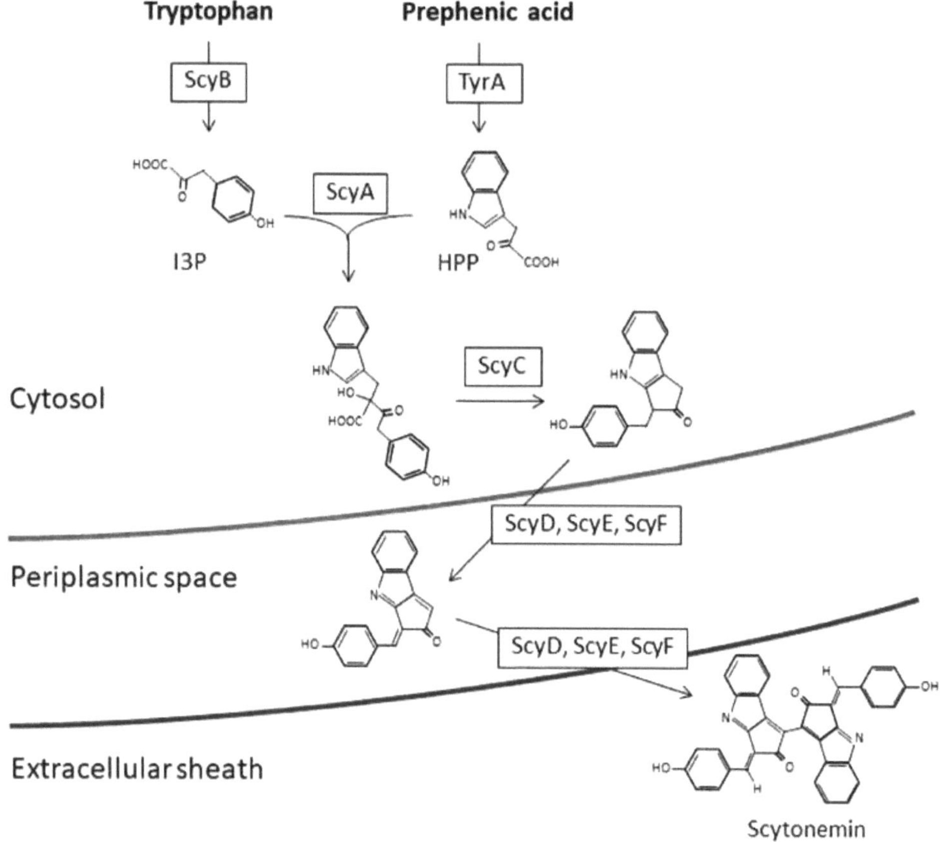

Fig. (2). Scytonemin biosynthetic pathway. See the text for an explanation.

REFERENCES

[1] Soule, T.; Stout, V.; Swingley, W. D.; Meeks, J. C.; Garcia-Pichel, F. Molecular genetics and genomic analysis of scytonemin biosynthesis in Nostoc punctiforme ATCC 29133. *J Bacteriol* **2007**, *189* (12), 4465-4472. DOI: 10.1128/JB.01816-06.

[2] Soule, T.; Palmer, K.; Gao, Q.; Potrafka, R. M.; Stout, V.; Garcia-Pichel, F. A comparative genomics approach to understanding the biosynthesis of the sunscreen scytonemin in cyanobacteria. *BMC Genomics* **2009**, *10*, 336. DOI: 10.1186/1471-2164-10-336.

[3] Balskus, E. P.; Walsh, C. T. Investigating the initial steps in the biosynthesis of cyanobacterial sunscreen scytonemin. *J Am Chem Soc* **2008**, *130* (46), 15260-15261. DOI: 10.1021/ja807192u.

[4] Balskus, E. P.; Walsh, C. T. An enzymatic cyclopentyl[b]indole formation involved in scytonemin biosynthesis. *J Am Chem Soc* **2009**, *131* (41), 14648-14649. DOI: 10.1021/ja906752u.

[5] Naurin, S.; Bennett, J.; Videau, P.; Philmus, B.; Soule, T. The response regulator Npun_F1278 is essential for scytonemin biosynthesis in the cyanobacterium Nostoc punctiforme ATCC 29133. *J Phycol* **2016**, *52* (4), 564-571. DOI: 10.1111/jpy.12414.

[44] Soule, T.; Stout, V.; Swingley, W. D.; Meeks, J. C.; Garcia-Pichel, F. Molecular genetics and genomic analysis of scytonemin biosynthesis in Nostoc punctiforme ATCC 29133. J Bacteriol 2007, 189 (12), 4465-4472. DOI: 10.1128/JB.01816-06.

SUBJECT INDEX

A

B

C

Hakuto Kageyama

D

E

F

G

H

I

J

K

L

M

R

S

T

U

W

www.ingramcontent.com/pod-product-compliance
Lightning Source LLC
LaVergne TN
LVHW070122110826
845147LV00002B/168